Tracking Apollo to the Moon

Springer

London
Berlin
Heidelberg
New York
Barcelona
Hong Kong
Milan
Paris
Singapore
Tokyo

Hamish Lindsay

Tracking Apollo to the Moon

With 250 Figures
including 106 Colour Plates

Springer

Hamish Lindsay
High Court of Australia, Parkes Place, Canberra, ACT 2600,
Australia

Endpapers: 111 km above the crater Schubert B in Smyth's Sea, the Apollo 16 Lunar Module *Orion* slips across the tortured surface of the Moon to rendezvous with the Command Module *Casper* for the trip back to Earth.
Frontispiece: From 8 km away, the slopes of South Massif dwarf the lonely Apollo 17 Lunar Module *Challenger* sitting in the valley of Taurus-Littrow.

ISBN 1-85233-212-3 Springer-Verlag London Berlin Heidelberg

British Library Cataloguing in Publication Data
Lindsay, Hamish
    Tracking Apollo to the moon
    1. Apollo 11 (Spacecraft)   2. Project Apollo (U.S.)   3. Space flight to the moon
    I. Title
    629.4'54'0973
    ISBN 1852332123

Library of Congress Cataloguing-in-Publication Data
Lindsay, Hamish.
    Tracking Apollo to the moon / Hamish Lindsay.
      p. cm.
    Includes index.
    ISBN 1-85233-212-3 (alk. paper)
    1. Project Apollo (U.S) – History. I. Title.
TL789.8.U6 A5467 2001
629.45'4'0973–dc21   00-058348

Typeset by EXPO Holdings, Malaysia
Printed and bound by Kyodo Printing Co. (S'pore) Pte. Ltd, Singapore
58/3830-543210   Printed on acid-free paper   SPIN 10742612

Dedication

This book is dedicated to my wife and our three children, grandchildren, and the upcoming generations so they can share in the exciting moments of mankind's greatest technological feat of leaving the Planet Earth for the first time and landing on our nearest neighbour, the Moon.

It is also dedicated to all those people whose contribution, however small, led to the incredible success of this feat. It is to remind us of the important part played by the support people on the ground – the Mission Control Center at Houston, Texas, the Kennedy Launch Center at Cape Canaveral, the Goddard Space Flight Center, Maryland, the Tracking Stations around the world, the Recovery Forces, and the many smaller centers and contractors whose people worked alongside the astronauts to keep the missions going to the end and bring them safely back home.

The Last Frontier

The wise man looks into space and does not regard the small as too little, nor the great as too big, for he knows there is no limit to dimensions.

Lao Tse, Ancient Chinese Philosopher

I do not for a moment believe that the spiritual well-being of our nation depends primarily on a space programme. I am sure that we could, as a nation, attain great spiritual reassurance from rebuilding our cities or distributing our farm products better. But I also believe that there are moments in history when challenges occur of such a compelling nature that to miss them is to miss the meaning of an epoch.Space is such a challenge.

Congressional Testimony by James A. Michener

Until Apollo 11 there was always a new frontier somewhere to beckon the inquisitive, exploring, often plundering human. We have now explored our planet from pole to pole, from the bottom of the seas to the highest mountains. Over the ages many have looked beyond the skies to the stars and have had the dream of travelling to other worlds, and walking on another planet. Countless stories have been written on aliens and the inhabitants of the other planets and fictional encounters with them. In the foreseeable future, the generations that follow us after Apollo 11 will never know the unique thrill of arriving in a new place, finding new people, new flora and fauna, new minerals, a new land. They will never experience the wonder, excitement and fear of the unknown that a new land, a new frontier can bring. We already know what Mars and Venus have to offer from our mapping orbiter and sampling lander spacecraft. We have crossed the last frontier in our solar system, in fact the vanguard of our spacecraft are already entering the intragalactic void.

If there is one thing the Apollo missions and the planetary explorers have shown us, it is there is nowhere else for us to go and live normally in our solar system.

It was not a dream that finally put a man on the moon, it was two powerful forces that came together at the one time – politics to provide the reason, drive, and finance, and science and engineering to provide the technology and vehicles. It had, however, taken visionaries from many times and many lands to develop the paving stones that built the track to the Moon. This is the story of putting those pavers down, and the part that the tracking stations played in the final scenes that led to man walking on the moon, and perhaps in time to come, to travel into the vast, incomprehensible reaches of the Milky Way Galaxy. One day Mankinds very survival may well depend on interstellar travel.

Hamish Lindsay
Canberra

Acknowledgements

"Tell it just like it was," is the theme of this book, and this philosophy has been followed, first hand accounts used wherever possible.

First of all I would like to thank all the many participants from around the world who helped to complete the story. Their names appear throughout the text, many interrupting very busy lives to add their bit. In particular the astronauts, who took some tracking down, but were very helpful in answering my questions, and have added that atmosphere to help "tell it like it was."

I would like to ackowledge the help of the following:

Lex Howard, Marshal of the High Court of Australia, who sparked the idea off over a game of table tennis, and supported the project from the beginning; Margaret Persinger at the Kennedy Space Center and Michael Gentry from the Johnson Space Center, Houston, who, with Becky Friday and Gloria Sanchez, provided most of the amazing images I requested from the ten million odd images held by NASA; G. Ted Ankrum and Dr Miriam Baltuck, NASA Senior Scientific Representatives in Australia; Michael Dinn, past Station Director, and staff of the Tidbinbilla Tracking Station, Canberra; Kerrie Doherty of the Power House Museum in Sydney; David Taylor of West Australian Newspapers, Perth; Dr Douglas Milne and John Masterton of the CSIRO for the Parkes input; Eileen Hawley, Public Relations Officer of the Medical team at the Johnson Space Center; Frank Winter, Curator of Rocketry at the Air and Space Museum, Smithsonian Institute, Washington; Linda Henry for her work in locating the items from American Society of Experimental Test Pilots, Lancaster, California; William Wood Jr late of the Bendix Field Engineering Company, for the Goldstone Tracking Station memories and photographs, and checking the manuscript; Joan Westbrook for the use of her scrapbook on the Mercury Project; Ed von Renouard, television technician at Honeysuckle Creek who was instrumental in arranging publication of the book. H.L.

Picture Credits

All photographs are by **NASA** except the following:

Author: Figures 1.1, 2.4, 2.5, 2.6, 3.3, 3.4, 3.5, 3.14, 4.2, 4.23, 4.24, 4.25, 4.27, 4.28, 4.29, 4.40, 4.42, 5.6, 6.14, 6.24, 6.29, 6.30, 6.43, 6.49, 7.17, A.2
Novosti: Figures 1.2, 2.10, 2.11, 2.12, 2.13, 2.34, 3.7
Deutsches Museum: Figure 1.4
Edwards Air Force Base, California: Figure 1.5
Spaceflight: Figure 2.11
Discovery: Figure 2.17
Lunar and Planetary Institute: Figure A.1
West Australian Newspapers: Figures 2.26, 2.28, 2.31, 2.32, 2.34
William Wood, Goldstone: Figures 4.18, 4.19, 4.20, 4.21, 4.22, 5.4
Australian News & Information Bureau: Figures 4.26, 4.30
John Masterson of the Australian Commonwealth Scientific & Industrial Research Organisation (CSIRO): Figures 4.51, 4.52, 4.53
Ed von Renouard, Honeysuckle Creek: Figures 5.7, 6.32
Peter Cohn, Honeysuckle Creek: Figure 6.23

Cartoons by Paul Rigby, courtesy of West Australian Newspapers

Foreword

One of the wonderful aspects of the US Manned Spaceflight Program was the opportunity for people around the entire globe to participate in one of man's greatest adventures. As we laid out the plans for flying the first manned spaceflight program, it was obvious that we would require extensive operations around the earth. One of the most challenging features of this plan was to build a world-wide network of tracking stations to provide communications with the orbiting spacecraft. At the time, about 1958 and 59, the construction of these facilities, in what turned out to be some very interesting pieces of geography, was a tremendous task.

Christopher C. Kraft, Jr.

Australia is located roughly 180 degrees longitude from the launch site, Cape Canaveral, and so occupied not only a unique position but a very critical one. Determining the position of the spacecraft as it traversed the Australian continent was critical to the orbit determination. This set of parameters was necessary to properly manage the entire operation. Such things as the time of retrofire, paramount to recovery of the crew, and the information required for signal acquisition at each of the tracking sites around the world are but two examples. Also, because the status of the astronaut and the spaceship were extremely critical to the decision-making process, the stations down under provided vital data to evaluate the progress and to allow the flight control team to manage the problems that inevitably developed.

Over the course of time, I knew and worked with a number of the people in Australia who contributed to our programs. They varied from technicians to communicators to managers in the Defense Department to Senators. I am very pleased that one of the people from Australia who lived in the times has chosen to write about the space odyssey of the 20th century.

Hamish Lindsay has done a marvelous job of telling the story of manned spaced flight.

He has given his readers a sort of encyclopedia of the beginning of man's quest for flight into space. This is followed with a splendid description of the real time operations of all of the major missions. As one who lived through the Camelot period of space in the 60s and knows the trauma we all endured, I am greatly impressed with the detail and authenticity of the stories that Hamish so vividly tells. For example, as I read Hamish's account of the Apollo 11 and 13 missions, I could again imagine myself back in mission control reliving some of the finest moments of my life.

Those of you who are fortunate to read Hamish Lindsay's account will be much richer for it.

Christopher C. Kraft, Jr.
NASA's First Flight Director
and Retired Director of the
NASA Johnson Space Center.
September 22, 2000

The Spirit of Apollo

In the next twenty centuries, the Age of Aquarius of the great year, the age for which our young people have such high hopes, humanity may begin to understand its most baffling mystery – where are we going?

The earth is, in fact, travelling many thousands of miles per hour in the direction of the constellation Hercules – to some unknown destination in the Cosmos. Man must understand his Universe in order to understand his destiny.

Mystery, however, is a very necessary ingredient in our lives. Mystery creates wonder, and wonder is the basis for man's desire to understand. Who knows what mysteries will be solved in our lifetime, and what new riddles will become the challenge of the new generations?

Neil A. Armstrong

Science has not mastered prophesy. We predict too much for the next year, yet far too little for the next ten. Responding to challenge is one of democracy's great strengths. Our successes in space lead us to hope that this strength can be used in the next decade in the solution of many of our planet's problems.

Several weeks ago I enjoyed the warmth of reflection on the true meanings of the spirit of Apollo. I stood in the highlands of this nation, near the Continental Divide, introducing my sons to the wonders of nature, and the pleasures of looking for deer and elk.

In their enthusiasm for the view they frequently stumbled on the rocky trails, but when they looked only to their footing they did not see the elk. To those of you who have advocated looking high we owe our sincere gratitude, for you have granted us the opportunity to see some of the grandest views of the Creator.

To those of you who have been our honest critics, we also thank, for you have reminded us that we dare not forget to watch the trail. We carried on Apollo 11 two flags of this Union that had flown over the Capitol, one over the House of Representatives, one over the Senate. It is our privilege to return them now in these Halls which exemplify man's highest purpose – to serve one's fellow man.

We thank you, on behalf of all the men of Apollo, for giving us the privileges of joining you in serving – for all mankind.

Neil A. Armstrong
Part of an address to a joint session of Congress on 16 September 1969

Contents

1 Origins

The longest journey begins with a single step.
Chinese proverb

We know the last step of our journey along the track to the Moon's surface down to a second, but who can isolate the first step? It winds back into antiquity, to people and places not even recorded. This book takes the reader briefly down the track from the earliest known beginnings to those heady days of Apollo, when walking on the Moon went from the impossible to part of our everyday life here on Earth in a human lifetime. It is interesting to note that the Moon was first seen through a telescope in 1609, had its first contact with humans in 1959, received the first visitor in 1969, the Apollo era ended with the death of Skylab in 1979, and the first human was buried on its surface in 1999.

As our nearest neighbour, it has affected our lives on Earth since mankind's earliest times, inspiring more superstition than any other celestial body, often as an omen of evil. The very word "lunatic" is derived from the Latin word for moon. Shedding its silver light over the Earth during the night, it has shone down on all the forms of life that have tried to exist on this Earth.

The Sun is always the same, but the Moon's appearance to us on Earth changes – waxing, waning, disappearing, then returning after three nights. Sometimes it peers down through wild and ragged storm clouds; sometimes it casts an eerie silver light to make sinister shadows in a still night, but can also sit at the end of a romantic path of shimmering gold on water framed by tropical palm trees. The Moon can rise red from smoke, or blue or green from volcanic dust, and appears to change its size as it rises.

Because of these observations and the many natural consequences on Earth from the light and gravity of the Moon, myths about the Moon tended to be dramatic and mysterious. Speculation on this object in our sky bred countless stories of what the Moon was doing to us: the light of the new Moon shining on your purse will keep you poor; cut your toenails during the new Moon so they won't become ingrown; make wine in

◀ **Figure 1.1.** The first step off the Earth. By stoking up a straw fire, Jean de Rozier and the Marquis d'Arlandes flew 8.8 kilometres over Paris in November 1783.

the Dark of the Moon; calves weaned during a Full Moon produce cows with the best milking qualities; sleeping in its light can cause blindness or madness; it has magic rays that dull knives; it makes women fall pregnant; it inspired fertility rites among Asians, human sacrifices among Celts – the list is endless. The Moon also provided gods for the Babylonians and Egyptians; and provided time and calendar references for many civilisations.

Then there is the Light and Dark of the Moon. Popular opinion generally has the Dark of the Moon as the period from Full Moon to the New Moon when it is waning, and the Light of the Moon is from the New Moon to the Full Moon when it is waxing.

There were many Moon gods among the ancient civilisations around the world. Excavations in the Tigris/Euphrates valley show that around 4,000 BC the Sumerians looked upon the Moon as the god Nanna, the chief god. Many temples were built for the god Nanna in the big cities of Sumer.

As the Moon's phases generally followed the menstrual period of women the gods became goddesses. There was a belief that when the Moon was full it was unlucky to work or cook on this "evil" day. Eventually, the Babylonians began observing a shabbattu, or Sabbath, at each quarter of the moon. The Jews adopted the custom and passed it on to Christianity.

Stonehenge has been mentioned as one of the earliest signs of the Moon being recorded by humans. Early graves have offered up bones marked with primitive indications of the various phases of the moon.

By 3,000 BC the Babylonians were able to calculate the movements of the Sun, Moon and planets in advance, and produced lunar calendars, starting their year with the first new Moon following the spring equinox. The Egyptians' Moon god was Thoth, who held sway over darkness, the underworld, and all the enemies of Re, the Sun god.

How our impression of the Moon has changed: from a god of mystery, a god that ruled the lives of people in past civilisations, to a barren, lifeless chunk of basalt littered with 100,000 craters just on the side visible to us, and the corpses of Russian and American spacecraft scattered over its surface. Rubbish dumps of cameras, boots, backpacks, and refuse litter the Apollo landing sites.

Will our Moon mean something different to future generations of humans?

Chronology of Events Leading to Lunar Exploration

2000 BC An Early Attempt to Reach Space

Mankind has been trying to reach the stars for a long time. One of the earliest known attempts to get into space was from an ancient Chinese legend dating from around 2,000 BC. Wan Hu, a rich mandarin, built a "space vessel" of kites propelled by 47 rockets. At his command 47 slaves leapt forward to light each of the 47 rockets, but a misfiring rocket caused an explosion, destroyed the machine, and the mandarin was never seen again.

160 AD The First Fictional Story Describing a Journey to the Moon

This was written by the Greek Lucian of Samosata. His book *True History* described a journey to the Moon. Sea voyagers beyond the Pillars of Hercules (the Straits of Gibraltar) were lifted up from the sea in a fierce Atlantic whirlwind and after an eight-day voyage came upon the Moon, a great country in the air, like a shining island.

On the Moon the sailors watched a war between Moon men and invaders from the Sun. Lucian described flights to the Moon, beyond the Sun, and to cities between the stars.

1040 The Origin of the First Rockets

The earliest reference to any kind of a device resembling a rocket is from China around 3 BC when bamboo tubes were filled with saltpetre, sulphur, and charcoal. Tossed into ceremonial fires to create explosions and noise, some would shoot out of the fire, propelled by the burning gunpowder. This was probably how the original concept of using rockets as a weapon was conceived. The recipe for gunpowder was first written down by the Chinese in 1040 AD, and later used in fire-arrows by the Chin Tartars in 1232 for fighting off Mongol assaults. A well known case was the siege of K'ai-fung-fu (now Peking) when the town fought off 30,000 Mongols with fire-arrows.[1]

1609 The First Telescope

Until the time of Galileo Galilei (1564–1642) no one could see any more of the Moon and stars other than by looking up into the open sky with the naked eye. In 1590 Italian opticians put together a crude form of telescope that could magnify an image up to three times.

When he heard about this Galileo made himself a telescope that he called "Old Discoverer". With a magnification of 33, he used it to begin the first serious observations of the heavens, studying the Sun to find sunspots, and discovering the mountains, craters, and librations of the moon, four of Jupiter's satellites, the first forms of rings around Saturn, and the planetary nature and phases of Venus.

Galileo estimated the heights of the lunar mountains from the shadows they cast and told the sceptics to look for themselves, then laughed uproariously when they claimed the mountains were not only impious, but did not exist. Although Galileo drew sketches of his observations of the features of the moon, they are not recognisable.

1651 The First Recognisable Maps of the Moon's Surface

These were drawn by Giovanni Riccioli. His *Almagestum novum*, published in 1651, gave features visible on the surface names of famous astronomers and philosophers. He called the dark areas seas and oceans; some of his romantic

Johannes Kepler (1571–1630)

Kepler was one of the founders of modern astronomy. He discovered the three basic laws which all planets and objects in space follow.

In a letter to Galileo, Kepler was the first person known to use the word "satellite". He discovered that the Moon causes the tides to rise and fall on Earth. He also wrote science fiction. In one of his stories called *Somnium Sive Astronomia Lunaris*, published in 1634 after his death, he described the launch from Earth "most uncomfortable and dangerous, for the traveller is torn aloft as if blown up by gunpowder." He explained the bitter cold and airlessness of space, discussed weightlessness, and even suggested the equivalent of reverse thrust to land gently on the moon.

Although Galileo and Kepler corresponded with each other, it seems they were not familiar with what the other was doing, so it was left to Isaac Newton to combine the fundamental work of these two great men to formulate the Law of Universal Gravitation.

names have remained to this day, such as *The Sea of Rains* and the *Ocean of Storms*. By the middle of the nineteenth century the absence of clouds was noticed and most of the main features of the near side had been identified.

1687 Isaac Newton (1642–1727)

Without doubt, Newton is one of the greatest scientists of all time. Among his many works he figured out the theory of gravitation and built the first reflecting telescope.

In 1687, Newton took one of the early steps along the track to the Moon by offering the first scientific principles of how all rockets operate, and recognised it as the only means of propulsion in space, in his third law of motion.

1783 The First Sustained Flight in the Air

Late in 1782, the year Britain lost America after the War of Independence, the Montgolfier brothers, sons of a paper manufacturer in the town of Annonay, near the centre of France, discovered that an inverted paper bag would float up if filled with hot air, which led to making the world's first hot-air balloon. Within a year, on 21 November 1783 at around 2:00 pm, Jean Pilâtre de Rozier (1756–1785) rose into the air above Paris under a Montgolfier balloon. With the Marquis d'Arlandes helping him stoke the straw fire, they became the first humans to experience the wonders of sustained flight and look down upon the Earth beneath (see Figure 1.1). They covered about 9 kilometres in 25 minutes.[2]

1865 Jules Verne (1828–1905)

Jules Verne is considered by many to be the founder of modern science fiction. He wrote *From the Earth to the Moon*, the first story of a trip to the Moon based on scientific principles, using calculations by his brother-in-law, a professor of astronomy. This Frenchman correctly foresaw an escape velocity of 40,200 kilometres per hour, weightlessness, collisions with meteorites, and steering by the use of firework rockets. His story also launched the mooncraft called *Columbiad* from Florida, which splashed down in the Pacific to be picked up by an American corvette, thus anticipating the flight of Apollo 11 that took place 104 years later! His story was the inspiration of the many space visionaries that followed, even helping Michael Collins to make up his mind to name the Apollo 11 Command Module *Columbia*. His grandson, retired Judge Jean Jules Verne, watching Apollo 8 leaving the launch pad, remembered his grandfather telling him, "I know that you will see men go to the moon, and you will be able to measure the accuracy of the images that I created."

1891 Grove Gilbert (1843–1918)

Gilbert is regarded by some as the greatest geologist that ever lived. With the formation of the United States Geological Survey (USGS) in 1879 he was made one of its six senior geologists. While Chief geologist of the USGS in 1891, he was considering the origins of the lunar craters and spent some time studying Coon Mountain in Arizona to determine whether it was created by volcanic action or was an impact crater.

In August, September and October of 1892 he spent 18 nights peering at the Moon through a 67cm telescope at the American Naval Observatory in Washington and presented the results in a paper called "The Moon's Face".

To explain the features he saw, Gilbert started with two working hypotheses – impact and volcanic – favouring the former as his observations seemed to lean more and more to the impact theory. Although it may have seemed of little importance at the time,

Gilbert's observations, sketches, and descriptions of the lunar surface and its craters were so remarkable for their accuracy and far-sighted interpretations that they would be accepted in any book on the Moon for generations to come, even after the Apollo excursions to the moon.[14]

1898 Konstantin Eduardovich Tsiolkovsky (1857–1935)

The self-taught Russian school teacher calculated the mathematical laws of rocket motion, and began to publish scores of articles about space travel, adding up to about 100 titles dealing with geology, cosmogony (a theory of the origins of the universe), aerodynamics and astronautics. In 1903 he published his classic *Exploration of Space with Reactive Devices*. In 1935 he published a book in Moscow called *On the Moon*.

He produced the design of a three-deck, manned spacecraft that included bathtubs of water so the cosmonauts could survive the gravitational forces of launch and re-entry. He suggested multi-stage rockets he called "rocket trains", and it was his descriptions of Earth satellites, liquid fuel rockets, space suits, solar energy, closed life-cycle systems, and eventual colonisation of the solar system that spurred the Russian scientists on to their major achievements after the Second World War.

He wrote, "Earth is the cradle of the mind, but one cannot live in the cradle forever," and the very prophetic, "For forty years I have been working on the reactive engine, and I thought that a journey to Mars will begin in hundreds of years. But time perspectives change. I believe that many of you will witness the first trans-atmospheric journey."[2]

Yuri Gagarin, the first man in space, wrote in the Guest of Honour book at Kaluga: "For us cosmonauts, the prophetic words of Tsiolkovsky on the opening up of outer space will always be our program, will always summon us forward."

Figure 1.2. Konstantin Tsiolkovsky: the amazing visions of this deaf schoolteacher spurred the Russian space programme to greater effort.

1903 Wilbur and Orville Wright (1867–1912 and 1871–1948)

The two brothers set the pace of the 20th Century by taking to the air in a powered aircraft on Thursday 17 December 1903. The day dawned with a cold, strong northerly wind blowing over ice glistening on the rain puddles. Dressed in a dark suit, stiff collar, necktie, and cap, Orville Wright braved the biting wind to guide their *Flyer* to a flight of 12 seconds, then after three more attempts finished with a best flight of 260 metres in 59 seconds.

The aviation industry was born.[2]

1915 The American National Advisory Committee for Aeronautics (NACA) was Created

Only ten years after the Wright brothers introduced the world to powered flight,

Europe was facing the First World War. Remote from the growing malevolence across the Atlantic, the United States was being left behind the rapid advances of the European aviation industry. Certain American leaders, seeing their technological advantage withering away, began to talk about restoring this advantage. Charles Walcott, then secretary of the Smithsonian Institution in Washington, began lobbying for a government agency to conduct research and experiments in aeronautics, finally convincing President Woodrow Wilson to sign a law establishing an Advisory Committee for Aeronautics on 3 March 1915.

A sum of $5,000 was allocated for the first year's operations. Twelve honorary members were appointed by the President from the leading military and scientific fields. As soon as they took office their investigations found that aviation in the United States at the time was generally regarded as a daredevil sport enjoyed by a few wealthy young men.

Under wise leadership and with skilful researchers NACA was soon helping the fledgling aviation industry to move forward. By 1930 NACA was a respected world leader in aeronautics with research centres spread across the country.[3]

1926 Robert Goddard (1882–1945) Launches the World's First Liquid Fuel Rocket

This was achieved on 16 March 1926 and was the embryo of the mighty Apollo Saturn V rocket. Lit by holding a paraffin blowtorch on the end of a pole, the rocket streaked away at 96 kilometres per hour to bury itself in the ground 56 metres away. At the time the world did not know of this momentous event as not a word was published in the popular press. Often referred to as the crazy Moon Professor, Goddard was given a hard time by his contemporaries, so he avoided publicity and put his thoughts of manned and unmanned spaceflight, navigation in space, and solar powered engines in a file labelled "Formulae for silvering mirrors" and locked them away in a cabinet. Working away on

Figure 1.3. Dr Robert Goddard in his workshop in New Mexico during the late 1930s. He fired the first liquid fuel rocket – an embryo of the later mighty Saturn V.

his rocket he would say, "I'm not trying to hit the Moon – I'm just trying to get this one off the ground."

It wasn't until the middle 1950s that the American rocket scientists discovered that it was virtually impossible to construct a rocket or launch a satellite without acknowledging the work of Goddard. Von Braun admitted, "Until 1936, Goddard was ahead of us all."

1928 Fritz Stamer

Stamer flew the world's first rocket-propelled aircraft, flying 1,500 metres in 80 seconds on the third attempt on 11 June 1928. On the fourth attempt, Stamer lit the fuse of one of two rockets and his Rhön-Rossitten-Gesellschaft *Ente* (German for "duck") glider shot into the air. Stamer describes the experience, "The take off went without a hitch. The first rocket was burning and I had really become accustomed to the very loud hissing of the jet flame spurting out of the nozzle, when about three seconds after ignition there was an ear splitting explosion... the entire aircraft was burning away merrily, and judging by the violence of the explosion, a few things contributing to its stability must have suffered damage too. I was particularly concerned about the wing suspension. I decided not to force the bird down vertically, although in doing so the flames would be pushed back to the rear, but to let it glide down carefully so as not to break up in the air. I was further comforted by the thought that the second rocket was there behind me in the fire, likely to go off, one way or another, at any moment.

Moreover, under my seat it was becoming first pleasantly, but then obtrusively warm. Fist sized chunks of powder from the exploding rocket had come flying in all directions into the machine and had set fire to it. One such chunk was now appropriately situated under the thin plywood seat. At last I grounded the machine. I made the finest landing and thereby possibly coming into closer contact with the second rocket. This was likely to go off at any moment as it was, and things would be in a bad way if I happened to crash right onto it. No sooner had the machine stopped than I had already climbed out of it. I saw the ignition wire burning on the iron rocket casing and I tried to tear it away. But it was already too late. The second rocket ignited, but it fortunately burned out in the proper manner, despite the intense heating of the steel jacket. If it, too, had exploded my prospects would not have been very bright. Now I wriggled about in the wet grass in order to extinguish and cool my smouldering posterior. My need to fly with powder rockets was temporarily satisfied...!"[5]

Stamer just lived to witness the landing on the moon, dying on 20 December 1969.

1929 Hermann Oberth (1894–1989)

Oberth was born on June 25 1894 in Hermannstadt, Transylvania, and became the third of the three recognised space visionaries who inspired his fellow countrymen to tackle the new and unknown arena of space. He had dreamed of space flight from being a child.

He came across a copy of Jules Verne's book and became enthralled in the story. Before he was thirty years old he had figured out ways to go up into space. In 1923 he published a slim volume on the theoretical aspects and mathematics of space travel, *Die Rakete zu den Planetenraumen* (*The Rocket into Interplanetary Space*). In it he also discussed a large multi-manned space station.

Then in 1929, Oberth published one of the first recognised "bibles" of astronautics, *Wege zu Raumschiffahrt* (*Ways to Space Travel*), expanding on his first book and dealing with all the fundamental engineering and mathematical problems of spaceflight. This comprehensive work was mainly responsible for the creation of the *Verein Für Raumschiffahrt*, or the "Society for Space Travel", more commonly known as the VfR. It began in the back of a restaurant in Berlin in July 1927, soon attracting over 500 members, including a youthful Wernher von Braun.

Oberth became a German citizen just before moving to work with von Braun at Peenemünde.

He arrived in New York during July 1955 and joined von Braun's team at Huntsville, Alabama, to work on putting American satellites into space using principles he had proposed more than thirty years earlier. A theoretician rather than a practical scientist, his contribution to the American space programme appears to have been limited, though he did publish a book, *Man into Space*, in 1957.

Oberth was the only one of the three recognised original space visionaries to live long enough to see the launch of Apollo 11 and his dreams come true, dying on December 29 1989.

1936 The Jet Propulsion Laboratory (JPL) was Formed

On 31 October, students of the Guggenheim Aeronautical Laboratory, a part of the California Institute of Technology (Caltech), created the Jet Propulsion Laboratory. It began work at the foot of the steep San Gabriel Mountains at Pasadena, near Los Angeles, and during World War II developed rocket assisted take-off systems for aircraft as well as missiles for the US Army. On 3 December 1958, JPL was contracted to NASA, and began a comprehensive planetary exploration programme that included a worldwide tracking network and the Pioneer and Voyager spacecraft now beyond the far reaches of Solar system in intra-galactic space. [12,15]

1939–45 The Second World War Spawns the V-2 Rocket

The German Army first became interested in the rocket as a weapon because the Treaty of Versailles prohibited Germany from using heavy artillery but did not specify rockets. Goering and Himmler were enthusiastic supporters of rockets, but initially Hitler was not particularly impressed with the rocket's potential for military uses. Around 1936 a rocket development centre was established on the island of Usedom in the Baltic Sea and took its name from the nearby fishing village of Peenemünde. Hitler never set foot at Peenemünde, only paying a visit to the Kummersdorf plant in March 1939. After seeing some demonstrations and touring the establishment, he commented, "Es war doch gewaltig," ("It was great") as he left, but later, in September of that year, he said that he

---

Wernher von Braun (1912–1977)

Wernher von Braun was born in Wirsitz (now Wyrzysk), Poland, and worked for Oberth when he was a student at the Berlin Institute of Technology. He sent for Oberth's book while he was still a junior high school student but found, "… to my great dismay I couldn't understand it because it was full of maths, not my strong point at the time."

Von Braun left school the year the Apollo 11 astronauts were born and offered his services to Oberth when he visited Berlin. His first job was to help with an exhibition in a department store and stood behind a table eight hours a day answering questions from shoppers and telling them, "There is no question that we will fly people to the Moon one day."

Von Braun began experimenting with liquid fuel development as a student with the VfR group until 1932 when he began to work for the German Army under General Walter Dornberger at Kummersdorf, near Berlin.

Sergei Korolev
(1907–1966)

Russia's Chief Designer was born in the town of Zhitomir in the Ukraine. He always thought of flying and the great sky above. He was so enthralled by a sea-plane flight while a teenager that he decided to dedicate his life to astronautics. His first success was the Koktebel glider, which Vasily Stepanchonok took through the world's first glider loop in 1930. He devoured all Tsiolkovsky's writings and was lucky enough to meet him just after founding the Leningrad Len-GIRD, a group for the study of reactive motion.

In 1933 he joined the Jet Propulsion Scientific Research Institute in Moscow as its Deputy Head, bringing out a book *Rocket Flight in the Stratosphere* the following year. In 1938 he was imprisoned for a life sentence of hard labour for associating with spies until the Soviet government released him from the Magadan gold mines during World War II to return to aircraft designing.

In 1946 he joined a party of observers at Peenemünde, staying in von Braun's house, before returning to Russia to work on the V-2 and Russia's space programme.

did not think the V-2 would be of any use to Germany in the war. This non-committal view set back the V-2 development programme by years. Later, when he was shown a film of the first successful trials, he changed his mind and, after apologising to Dornberger, gave the V-2 programme top priority.

Following the success of the A.5 (Assembly-5) rocket developed by von Braun's group with a thrust of 1,500 kilograms lasting 45 seconds, they were given instructions to work on an accurately directed, long distance rocket with a maximum range of 250 km and capable of carrying a 1,000 kilograms warhead. This 14-metre-long rocket was identified as the A.4 and became the infamous V-2, the *Vergeltungswaffe*, or Vengeance Weapon.

1942 The V-2 Flies for the First Time

After two spectacular failures, the V-2's first successful firing was on 3 October 1942. Workers gathered around the launching area to watch as 25,402 kilograms of thrust for 63 seconds sent the sleek V-2 slowly, then with rapidly increasing speed up into a cloudless sky. Dornberger, von Braun, and Oberth held their breaths watching the black and white projectile disappear into the azure blue distance above – this time there was no explosion, no unexpected events. Cheers, shouts and laughter echoed around the buildings. Dornberger turned to von Braun and shouted exultantly, "Do you realise what we accomplished today? Today, the spaceship is born."

1945 The First Vertical Rocket Flight

This was by Lothar Siebert in February 1945. The German Luftwaffe launched Siebert in a NATTER (German for "Viper") Bachem Ba missile straight up into the atmosphere. At 152 metres the stubby winged rocket lost its cockpit cover, which probably broke the pilot's neck as his headrest was part of it, before reaching a height of 1,524 metres. It then turned over and smashed straight into the ground. Unfortunately we do not have a first hand account of this flight.[6]

1945 The End of the Second World War

As the war finally came to an end, the Russian politicians in the Kremlin suddenly became aware that the Americans had bases all around their perimeter, the atomic bomb, and long range bombers such as the B29. They realised that their traditional tanks and infantry, even their Air Force, would not be able to reach the continental United States. In fact, at the time, the Soviet Union did not even have a jet aircraft, let alone a rocket. To make matters worse, on May 22 1945, just one day before the Russians were due to arrive in Nordhausen, von Braun and over a hundred of the top rocket scientists had fled to the west, taking most of their technical data with them. Their archives, rocket engines, pristine V-2s, machines, tools, and materials were loaded onto 300 freight train carriages and taken to Antwerp where they were transferred to 16 Liberty ships bound for New Orleans. By the northern autumn of 1945 it had all been safely delivered to the recently established White Sands Proving Ground in New Mexico in an operation code named "Paper Clip", organised by Major General H.N. Toftoy, commanding general of the Army Ballistic Agency at Huntsville, Alabama.

The invading Russian army found ruins, partly completed projects, a few damaged and incomplete V-2s, and only a handful of top Peenemünde engineers and administrators - not a single leading German rocket designer.

1946 The First Man-made Object Enters Space

The German rocket scientists settled into the US Army White Sands Proving Ground with their wartime V-2s, and began to develop rockets for the US army. On 16 April 1946, four years after its first firing in Germany, the US Army launched its first V-2 rocket in America. Unfortunately, a fin fell off and the rocket crashed.

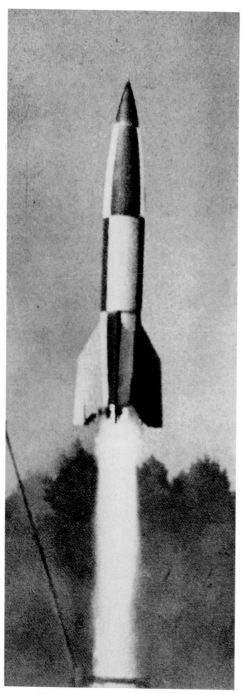

Figure 1.4. A German V-2 rocket.

As far as can be ascertained the Germans never actually entered space during the war. The first man-made object to pass the 122 kilometre mark, the Apollo definition of where space begins, and enter space appears to be White Sands V-2 #7, which reached a height of 133.5 kilometres on July 9 1946.[7]

Where does space actually begin?

Way back in 1895, Tsilokovsky estimated you had to be 320 km above the Earth to be above the atmosphere. However, it is impossible to define accurately the edge of the atmosphere as it varies in height due to seasonal pressure changes. Somewhere above 1000 km atmospheric gas molecules are negligible and the lighter molecules can escape the gravity of Earth.

The Fédération Aeronautique Internationale (FAI) in Paris defines space officially beginning at 100 km above sea level for their purposes, but it is at 160 km it is recognised that the laws of aerodynamics give way to the laws of astrodynamics.

From the Houston Mission Control Center's point of view during Apollo, space began at 122 km above sea level, as Flight Dynamics Officer (FDO) Jerry Bostick explains: "We used 400,000 feet. That's when you really start running into some resistance from the atmosphere. That's what we called 'entry interface'. From our point of view, if you were above 400,000 feet you were in space."

One of the original proposals to launch a spacecraft was put forward by the American Air Force's Project Rand (Research ANd Development), which submitted a report to Congress through House Report No 360 titled "Preliminary Design of an Experimental World Circling Space Ship" on 12 May 1946 that looked at the feasibility of building and launching an artificial satellite.

1946 Radio Astronomers begin Looking at the Moon

The first weak radio emissions from the Moon, generated by the Sun heating up the surface, were picked up by Robert Dicke during the Second World War at the Rad Lab in the Massachusetts Institute of Technology, though it wasn't until 1946 that the first radar signals bounced off the lunar surface were detected on Earth by operators in the United States and Hungary.

The following year the Australian Commonwealth Scientific and Industrial Research Organisation's (CSIRO) trio of Frank Kerr, Alex Shain, and Charles Higgins began the first scientific study of the lunar surface by radar. Using a powerful transmitter at the "Radio Australia" station at Shepparton in Victoria, and a receiving station at Hornsby in Sydney they began to analyse the echoes from the moon. In 1948 Jack Piddington and Harry Minnett began a series of observations on the bright and dark areas of the Moon. These showed that the surface temperature varied from –63°C to +17°C each month, but also found that the maximum temperature lagged behind the expected midday peak because the radio emission they were measuring came from layers beneath the surface which took longer to heat and cool. After studying heat conduction in solids they concluded that the lunar surface consisted of porous rock and gravel covered by a layer of fine dust with an average thickness of 2 cm. This conclusion was quite remarkable considering these results, later verified by the Apollo missions, came from only a small dish antenna.[34]

1947 The Russians begin Their Rocket Programme Behind the Iron Curtain

In 1946 Marshal Zhigarev, Commander in Chief of the Soviet Air Force, announced that the V-2 was useless against a distant enemy and they would have to develop their own reliable long-range rockets capable of targeting the American continent. On 15 March 1947 Marshal Stalin, the Soviet Premier, said, "The problem of developing transatlantic rockets is of extreme importance to us," and proposed the formation of a special State Commission to begin studying the development of long-range rockets.

The first Russian V-2 was launched from Kazakhstan on October 30 1947, 18 months after the first American V-2. It flew for 298 km and landed in the planned target area. A second in November destroyed itself.

In 1949 the Russians began researching the effects of spaceflight on animals by sending dogs in pressurised containers up to heights of 97 kilometres then later to 450 kilometres.[8]

America was already experimenting with sending primates into the atmosphere. In June 1948 at White Sands, V-2s called Albert and Albert II, followed by two more attempts, all ended with the animals surviving the flight but dying before the capsule was located.

1947 **The Sound Barrier is Broken by a Human**

On 14 October 1947 the silent desert morning air was shattered by four engines straining to drag the heavy bulk of a B-29 bomber into the sky over Muroc Dry Lake in California. Strapped under a wing was an experimental rocket aircraft, the Bell X-1 Glamorous Glennis. Captain Charles 'Chuck' Yeager settled into the tiny cockpit and Flight Engineer Jack Ridley dropped the hatch into place. Yeager locked it down with a shortened broomstick, wincing as he fought the pain of two broken ribs from a horse riding accident two days before. Yeager recounts, "All we had in the way of protective gear was a leather World War II helmet, with the top of a football helmet cut off and snapped onto the top for protection against banging your head against the top of the cockpit, a plain wool flying suit, and a leather A-2 jacket the same as we wore during World War II."[10]

Forty five minutes later he was dropped from the labouring B-29 at around 7,600 metres and lit all four rockets to power up at a 45° angle to break through the sound barrier at a speed of 1,127 kilometres per hour, or Mach 1.06. He climbed to a height of

Figure 1.5. Captain Charles Yeager beside his *Glamorous Glennis*.

13.106 kilometres and watched the sky turn a deep purple, and the stars and Moon appear with the bright Sun. On the ground the watchers lost sight of the aircraft but heard the sonic boom of man breaking the sound barrier for the first time.

In August 1950, General Hoyt Vandenberg, then Air Force Chief of Staff, said when presenting the X-1 to the Smithsonian Institution, "The Bell X-1 marked the end of the first great period of the air age, and the beginning of the second. In a few moments the subsonic period became history, and the supersonic period was born."

1949 **The World's First Government Department of Space Medicine was Established in February 1949**

Not as well publicised as their rocket colleagues, a group of German Luftwaffe high speed, high altitude, medical specialists were brought together at the Wright Patterson Air Force Base, near Dayton, Ohio, to begin work on the medical aspects of man in space. Included in the group was Hubertus Strughold, later to become known as the "Father of Space Medicine".[9]

1949 *Bumper* **Reaches Out into Space**

Project Bumper comprised a two stage rocket using the German V-2 as a first stage with a high altitude rocket developed by the California Institute of Technology, called a WAC Corporal, as the second stage. The V-2 lifted the WAC Corporal to 160 kilometres above the earth, and the Corporal then continued on to a record height of 393 kilometres at a speed of 8,288 kilometres per hour on 24 February 1949.[9]

The seventh *Bumper*, the main stage again a V-2, was the first attempt to launch a rocket from the new Long Range Proving Ground at Cape Canaveral in Florida, later to become the Kennedy Space Center. Owing to a valve malfunction it was replaced on 24 July 1950 with the eighth *Bumper* that became the first successful launch from this famous spot. In those days the launch centre was nothing more than gravel roads and sandbagged bunkers.

1954 **The Russians begin Plans to Enter Space**

The scene was now being set for the last stages of the assault on the Moon. The leading political players were now on centre stage and, in the wings, the engineering, medical and technology specialists were gathering their forces to support the politicians. The rocket had developed into a practical engine for launching spacecraft, and computers, though primitive, were able to do the work required. It has been estimated all the calculations required for a trip to the Moon would take a person 800 years – computers can do it in 30 seconds.

The seeds for Sputnik, the world's first Earth-orbiting satellite, were sown when Sergei Korolev, the Russian rocket mastermind, got together with the Soviet Premier, Nikita Khrushchev, and hatched the Russian space programme. Korolev, in those days only referred to publicly as "The Chief Designer" to conceal his real name, wanted to explore space, whereas Khrushchev wanted something to dazzle the world with the power and technical prowess of the Soviet Union. All the better if it was Russian cosmonauts stepping on the surface of the Moon. However, not all Russian politicians and scientists agreed with these proposals. In 1955 the USSR Academy of Sciences sent a letter to over a hundred scientists for their views about the use of artificial Earth satellites. The responses varied: some were not interested in such pure fantasy; others felt certain that satellites would not needed before the year 2000; yet others could see no practical uses for artificial satellites; but there were also a number of constructive suggestions. By the mid-1950s Russia was an advanced rocket power working on two programmes – missiles and spaceflight.

1955 Russia Constructs Its Own Launch Facility

In January the first construction workers arrived at the site of the proposed Soviet launching facility at Tyuratam, staying at the town of Zarya (renamed Leninsk in January 1958). The launch facility became known as Baikonur Space Centre, planned as a deception to mislead the West by referring to a town some 370 km from the City of Leninsk which had grown up around a rail head near the launch site.[8]

In the same year the Soviets had produced a multistage rocket design using full size mockups and drawings. They improved the original German V-2 design and assembled them into a cluster of 20 motors of 34,927 kilograms thrust each, which gave them the power to launch the heavily constructed sphere of the man-rated Vostok spacecraft.

On 26 April 1955 Moscow radio announced that the USSR planned to explore the Moon with a tank remotely controlled by radio, predicted trips by man into space in one to two years, and reported formation of a scientific team to devise a satellite able to circle the Earth.

In April 1956 the Russians produced the first official plans to enter space in 1958 with concepts of orbiting reconnaissance spacecraft leading to the possibility of manned flights to the moon. At least three groups, led by Korolev, Glushko, and Bondaryuk, began working on the design of a nuclear booster and Earth satellite, and a preliminary programme to get a man into space.

1956 The Americans Take a First Look at Manned Space Flight

In America the first manned flight studies were begun by the US Air Force as Project 7969, Task 27544, in March 1956. Called *A Manned Ballistic Rocket Research System*, it only lasted nine months before being filed away for lack of funds. It proposed to use the Atlas booster and the *Hustler*, later to be known as the *Agena* rocket in the Gemini programme.[9]

1957 The International Geophysical Year (IGY)

This was held from 1 July 1957 to 31 December 1958. For the first time in history the physical Earth was to be studied simultaneously from over 2,500 scientific stations by more than 5,000 scientists with the co-operation of 67 countries. As part of America's contribution to the International Geophysical Year, the American Air Force was to launch three spacecraft to the Moon, and the Army two, while the Army's Jet Propulsion Laboratory (JPL) in California was to design the spacecraft itself.

In 1954 the International Council of Scientific Unions decided that satellites in space should be included in the International Geophysical Year programme. This gave both the Russians and Americans a boost to their space activities, and on 15 April 1955 the Soviets announced that they would try and launch a satellite for the IGY. The Americans made their announcement on 29 July 1955. At the time the Russians worked away in secrecy, while the Americans publicly displayed their spectacular failures.[11]

1957 The World's First Artificial Satellite goes into Orbit

Then, suddenly, a new word hit the news bulletins around the world: "Sputnik".

On 4 October 1957, only 54 years after the first aircraft dragged itself into the air for 12 apprehensive seconds of powered flight, the shiny metal sphere of Sputnik 1 popped

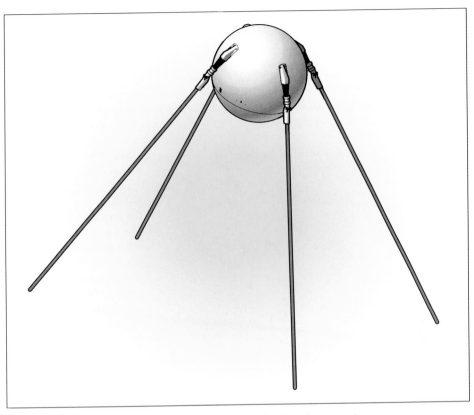

Figure 1.6. The Earth's first artificial satellite: Sputnik.

confidently out of the Earth's atmosphere to extend mankind's horizons to the limitless reaches of outer space and opened up a whole new set of challenging frontiers.

Although there had been references in the press to the Russians developing an artificial satellite, the West was taken by surprise, and the world's headlines screamed: "Red Moon is seen while the world listens to its signals"; "Bleep! and it whirls on"; and "Russia wins space race...."

The Russian scientists and engineers shed tears of joy, embraced each other, and sang the Russian Air Force anthem while above them Sputnik 1 orbited the Earth, easily visible as a bright star shooting across the sky in the evening.

Around the whole world "Sputnik", or "Travelling Companion", a name first used for an artificial satellite by Tsiolkovsky back in 1903, became a household sensation overnight, and Khrushchev's plan to show the glory of the Soviet Union's technological prowess to the people of the Earth was right on target. Everybody was talking about it. Radio hams around the world picked up its signal.

The author remembers: "One night I was walking down a dark Tasmanian country road and looked up to see the bright spot of light that was Sputnik streaking across the sky. It was the first time the people of Earth saw a man-made object in space."

This tiny 58 cm aluminium ball with four rod antennas, weighing all of 83.5 kilograms, riding one of Korolev's R-7 ICBM launch vehicles from Tyuratam, opened the door to space and the vast areas beyond. Sputnik 1 stayed up for three months, its only experiment a 1-watt radio transmitter sending signals on 20.005 and 40.002 MHz. It took

96 minutes 12 seconds to zoom around its initial orbit with a apogee of 941 km and perigee of 219 km. After about 1,400 orbits it finally sank back into the atmosphere to burn up on 4 January 1958.

1957 The American Reaction to Sputnik

At a news conference on 9 October, a miffed President Eisenhower declared, "From what they say they have put one small ball in the air and at this moment you don't have to fear the intelligence aspects of this."

Figure 1.7. The US Navy's Vanguard disintegrates.

Homer Newall, Associate Administrator of NASA, said, "In the United States many were taken aback by the intensity of the reaction. Hysteria was the term used by some writers, although that doubtless was too strong a word. Concern and apprehension were better descriptions. Especially in the matter of possible military applications there was concern, and many judged it unthinkable that the United States should allow any other power to get into a position to deny America the benefits and protection that a space capability might afford. A strong and quick response was deemed essential."[13]

One member of a Texas city council moved a motion to make it illegal to allow Sputnik to fly over the city, but there being no seconder, and there was no legal way they could stop it, the motion was dropped.

Dr William Pickering, the New Zealand-born Director of the Jet Propulsion Laboratory, already had a programme he called *Red Socks* ready to go to the Moon with nine spacecraft, but the Department of Defense wasn't interested. A new group called the Advanced Research Projects Agency had so many space proposals piled up in their 'IN' tray the Deputy Director was moved to comment, "It seemed to me that everybody in the country had come in with a proposal except Fanny Farmer Candy, and I expected them any minute."

The Agency felt unmanned flights to the Moon were the best way to beat the Russians. In March 1958 President Eisenhower approved proposals to send unmanned scientific spacecraft to explore the Moon.[15]

In the defence establishments of the West a fear grew that the Russians may already have unannounced and undetected reconnaissance satellites and weapons in orbit, and a crash programme began in America to try and find them. A $3.5 million project called SPASUR (Space Surveillance System), using Minitrack tracking stations was set up across the United States to try and locate unidentified objects in space.[12]

The Americans, complacently believing that their Vanguard rocket was going to lift the world's first satellite into space, were still preparing to get it off the ground when on 3 November 1957 the Russians sent the dog *Laika*, or "Little Barker", into space in Sputnik 2. The world's first biosatellite, it was promptly dubbed *Muttnik*. The animal proved living things could survive the rigours of launch and exist in space, before she was killed by an injection and burnt up in the atmosphere on April 14 1958.

The Soviet space surgeons chose dogs for space because they had collected considerable data on them over many years of research, and because their circulation and respiration is close to that of man. Other factors were their obedience and durability during long experiments. They were all female mongrels chosen from the Moscow area, with white hair so as to be seen and photographed easily.

Obviously, the Russians were preparing to send a man up and, now under heavy political pressure, the US Navy rushed their Vanguard preparations through only to see the three stage rocket rise a few centimetres, shudder, and dissolve into a bright ball of orange flame on the launch pad on 6 December 1957. Along with Congress and the space teams, a television audience of millions felt the stinging humiliation of failure, made all the more galling when they realised the Russians had success-fully lifted 508 kilograms into space, while the Americans couldn't handle a mere 1.5 kilograms.

1958 Explorer Became the First American Spacecraft to Reach Orbit

With the Navy's attempt a failure, the pressure was applied to von Braun and the Army's Jupiter C rocket using the Redstone booster. The Department of Defense had already rejected von Braun's offer to launch the *Orbiter* satellite the year before because the IGY was an international venture for peaceful scientific research, and as von Braun's Redstone had been developed for military purposes, it seemed inappropriate. But now, when von Braun and his team were given the go-ahead they promised to launch a satellite into space within 90 days and quickly prepared a satellite called *Explorer 1* for launching. Delayed by unusual storms in Florida and by high winds in the upper atmosphere, the 14-kilogram satellite was finally hurled up into orbit on the end of a JUNO 1 rocket at 10:48 pm on 31 January 1958. Watchers saw the upper stages begin to turn, until they were spinning at nearly 500 rpm, then, riding on top of a pink cloud of smoke and steam, it rose into the black sky. The rocket's glowing red trail pierced a cloud, then faded into the distance above their heads, as its roar drowned out sporadic applause and cheers. Rising to a maximum height of 2,531 km, *Explorer 1* made history by discovering the Van Allen radiation belts around the Earth with its Geiger counter, regarded by many as the most significant discovery of the IGY. During the battery's 112-day life, the satellite transmitted cosmic ray, micrometeorite and temperature data from instruments designed and built by Dr. William Pickering and his team at the Jet Propulsion Laboratory, Pasadena, California. Five of the first seven Explorer series satellites successfully reached Earth orbit, *Explorer 6* televising the first cloud cover picture of the Earth.[9]

1958 NACA begins Manned Space Flight Plans

On 29–31 January 1958 a conference was held at the Wright Patterson Air Force Base, Ohio, to review concepts for manned orbital vehicles. NACA formally presented two concepts, and contractors presented their proposals.

The Advanced Research Projects Agency then sent a letter to the US Air Force on 28 February 1958 telling it to get a man in space as soon as technology permitted.

In response to the above letter, the US Air Force began looking at manned space flight with a programme called MISS, "Man In Space Soonest", with a conference held at the Air Force Ballistic Division in Los Angeles on 10–12 March. Using a high-drag, no-lift, blunt shaped spacecraft with parachute landing, they planned to begin orbital flights about 1960, ending with landing a man on the Moon in 1965 at an estimated cost of $1.5 billion. In 1958 this programme, with technical support from NACA, was regarded as the most likely US manned space flight programme to get off the ground. These two agencies also began looking at tracking facilities that may be required for MISS.[9,15]

As some of these tracking stations would have to be built in foreign countries, military installations were out of the question at the time. This was one of the reasons why Congress wanted to form a civilian space agency to look after the American space programmes.

1958: 1 October The National Aeronautics and Space Administration (NASA) was Born

In January 1958 the American Rocket Society and the Rocket and Satellite Research Panel issued proposals for a National Space Establishment. President Eisenhower submitted a message to Congress calling for a special civilian Space Agency, using NACA as a nucleus, to conduct Federal aeronautic and space activities, and on 16 July 1958 Congress passed the National Aeronautics and Space Act of 1958. By 29 July it was signed into law by President Eisenhower's Executive Order 10783 with an annual budget of $100 million. On Tuesday afternoon, 30 September 1958 over 8,000 people left their NACA offices and returned the next morning to work for NASA, and immediately began to concentrate on developing a national coordinated space policy with the backing of Congress, away from the fractional attempts of the competing armed forces. Its first Administrator was Dr Thomas Keith Glennan, a distinguished administrator active in national and civic affairs. Within a week Project Mercury, to put a man in space, was approved.[3,9]

Figure 1.8. NASA's first Administrator, Dr Thomas Keith Glennan.

Increased pressure was put on the conservative Eisenhower Administration to do something about the Russian space menace, although many business associates were trying to hold back on the space budget. Eisenhower wrote to NASA saying he did not want the space budget to go over $2 billion per year. The rocket booster programme was under way, and they had approved the Mercury programme to put a man into space, but were reluctant to assign extra dollars for more ambitious plans.

1958 First Attempts to Reach the Moon

Generally known as the Pioneer Program, supporting the International Geophysical Year, America's early flights were catastrophic disasters. The first attempt to reach the Moon on 17 August 1958 left the launch pad, but 77 seconds later dissolved into a fireball when a bearing seized in the main engine. Nearly two months later the second effort failed when the spacecraft did not reach the proper escape velocity and fell back to be incinerated in the Earth's atmosphere. The Air Force's last attempt in early November fizzed when the third rocket stage never ignited, and the spacecraft tumbled back to Earth from 1,550 km.

The Army's first launch in December was planned to pass by the Moon into solar orbit, but also fell back to Earth after 38 hours aloft due to a short first stage burn.

1958 The American Space Task Group

This was formed on 5 November 1958 when 45 people gathered together as the nucleus of a new group that would eventually blossom into the Mission Control Center at Houston, later to become the Johnson Space Center. This youthful group was led by their oldest member, 45-year-old Robert Gilruth, and included a name that was to become a legend – Director of Flight Operations at Houston, Christopher Columbus Kraft, Jr, to become the Director of the Johnson Space Center in 1972.

Before World War II Gilruth had been involved in the design of the Laird Watt, the fastest aircraft in the world, and had already broken the sound barrier during the war with a bomb shaped device dropped from a B-29 bomber. At the time he was running a research establishment at Wallops Island, Chesapeake Bay, called PARD, Pilotless Aircraft Research Division. During the 1940s and early 1950s PARD was working with multistage rockets, so when the Space Task Group was formed, Gilruth was naturally chosen as its Director.[19]

1958 Project Mercury Announced to the Public

On 17 December 1958, the anniversary of the Wright brothers' flight at Kitty Hawk, Administrator Glennan revealed to the world that NASA was going to put a man into space with the new Project Mercury.

1959 Luna 1 Becomes the First Man-made Object to Leave the Earth

To further rile the Americans, on 2 January 1959 the Russians successfully launched *Mechta*, more commonly known as Luna 1, to become the first spacecraft to reach the 40,000 kilometre per hour speed required to break away from the Earth's gravity. A 363-kilogram sphere with four antennas and a probe, it flew within 6,000 km of the Moon, before going into solar orbit as an artificial planetoid. It takes 450 days to go around the Sun with a aphelion of 198 million kilometres and a perihelion of 147 million kilometres.[9]

A few weeks later the US Army did manage to get a rocket to behave as designed and their Pioneer 4 flew past the Moon at a distance of 60,000 kilometres.

1959 **NASA Decides to Explore the Moon**

During a train trip to Canada Nobel Laureate Geochemist Harold Urey (1893–1981) began reading a book called *The Face of the Moon* by astrophysicist Ralph Baldwin, and became totally absorbed in the Moon, seeing it as a probable source of primordial material from the formation of the Earth. On 29 October 1958 he addressed the Third Lunar and Planetary Exploration Colloquium, and offered his thoughts to the members present of: "the unique importance of the Moon for understanding the origin of the Earth, and the other planets."

In January 1959 Urey presented a series of lectures in Washington. These were heard favourably by the NASA administration, who quickly formed a Working Group on Lunar Exploration chaired by Robert Jastrow, not long appointed as leader of a theoretical space sciences group.[14]

Homer Newall, NASA Associate Administrator, said, "Immediately upon joining NASA, Jastrow busied himself with promoting space science. He joined forces with Harold Urey to agitate for an early start of a lunar programme."[13]

1959 **The Russians Land on the Moon and Photograph the Far Side**

The Russians always seemed to be able to anticipate the American plans and keep one step ahead. On 12 September 1959 they launched their second Luna spacecraft which managed to impact on the lunar surface at Latitude 29.1°N Longitude 0.00° in Palus Putredinis after a 33 hours 32 minutes flight to become the first man made object to land on the Moon. Nearly a month later, on 4 October, Luna 3 went behind the Moon and from 6,200 km above the surface took the first photographs of the other side with a 35 mm camera using two lenses of 200 and 500 mm focal length. The pictures were developed on board and transmitted using power from solar cells. Luna 3 lasted until April 1960. At that time the Russians were unquestionably leading the way into space and on to the Moon.

The Americans, however, were not fazed by the unending Russian successes or the sight of their own rockets' spectacular pyrotechnic failures and out of control spacecraft. Backed by an increasingly anxious Congress, they redoubled their efforts to catch up and continued planning more sophisticated spacecraft, better instruments, lunar landings, and lunar exploratory equipment.

1960 **A Moon Landing? President Kennedy Looks at His Options**

John F. Kennedy, his new government smarting over the Bay of Pigs and other

When is the Moon upside down?

A person facing the Moon standing anywhere in the northern hemisphere sees the north pole of the Moon as the top, whereas a person facing the Moon in the southern hemisphere sees the north pole as the bottom – 180 degrees opposite to the northern viewer.

Vince Ford, astronomer at the Mount Stromlo Observatory in Canberra, explains, "It is much easier to see this by doing a simple experiment. Go to one side of the room, say the southern side, and look at the ceiling light, noting which is the top, or north. Then go to the opposite side of the room and look again and you will see the light is now reversed – the top is now the bottom. Now imagine how confusing it is in the tropics where the Moon is sometimes north of you and sometimes south!"

Figure 1.12. Astronomer Gerard Kuiper (left) sharing in the excitement of the Ranger 8 close-up pictures with Raymond Heacock and Ewen Whitaker. Later, Whitaker accurately located Surveyor III's position for the Apollo 12 crew's visit.

2 The Mercury Project

> We who inhabit the Earth dwell like frogs at the bottom of a pool. Only if man could rise above the summit of the air could he behold the true Earth, the world in which we live.
>
> *Socrates, c. 410 BC*

Originally the project was referred to as the *Manned Satellite Program*. Abe Silverstein, Director of Space Flight Programs at NASA Headquarters, suggested the name *Project Mercury*, "because in Roman mythology, Mercury is the winged messenger of the gods." Robert Gilruth, then Project Manager wanted it called *Project Astronaut*, but the NASA hierarchy preferred *Mercury* because they wanted the emphasis placed on the machine rather than the man. The Olympian messenger Mercury was the most familiar of the Greek gods to the American people, rich in symbolic associations with advertising, automobiles, and metallurgy, so *Project Mercury* was adopted on 26 November 1958. Thomas Keith Glennan, the first NASA Administrator, revealed it to the world in a speech on 17 December 1958.[9]

Project Mercury had three main objectives:

1. To put a man in orbit around the Earth.
2. To see if a man could function in the environment of space.
3. To recover the man and spacecraft safely.

Using three main basic principles:

1. Simplest and most reliable approach.
2. Minimum of new developments.
3. Progressive build-up of tests.

Using these guidelines, regularly used to begin briefings at the time, the Americans took a scant three years to begin from scratch and develop the whole Mercury system to successfully launch an astronaut into space. To speed up the development most of the

◀ **Figure 2.1.** "You're on your way, José!" The Redstone rocket, with Alan Shepard in *Freedom 7* on top, clears the launch pad for a flawless sub-orbital hop to start America's manned flights into space.

equipment chosen was well tested and readily available from off the shelf and military sources.[9]

When the Mercury Project began in 1958, a manned Moon landing was still somewhere in the fuzzy future to the space visionaries at the time.

The Launch Vehicle

At the time the only rocket capable of lifting a manned spacecraft off the ground was von Braun's Army Redstone, a direct development of the German V-2 using fuel of liquid oxygen and ethyl alcohol, giving a thrust of 35,380 kilograms, and this was the vehicle used in the early Mercury flights, labelled MR missions. However, to get the Mercury spacecraft into orbit it needed the 165,564 kilograms thrust of the Atlas D model rocket. The Atlas, developed from the Convair MX-774 test vehicle first fired in July 1948, kept on blowing up with spectacular fireballs – right in front of the watching astronauts! "Are we are expected to fly this thing?" commented John Glenn, wryly, after watching one develop a crazy wobble and dissolve into a pyrotechnic inferno.

The Atlas, using liquid oxygen and liquid hydrocarbon fuel (RP-1), was modified for manned space flight, mainly by introducing an automatic impending catastrophic failure sensor, and a special programme to select quality components. In the event of a catastrophic failure of the booster rocket there was an escape rocket on top to blast the astro-

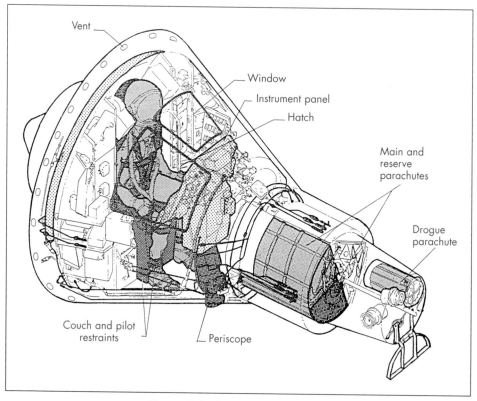

Figure 2.2. The Mercury spacecraft.

naut and his capsule to safety. First tested on 12 April 1959, this escape rocket was never used in an American manned mission.

There were 25 major flight tests during the Mercury Project, beginning with Little Joe 1 on 21 August 1959 and ending with MA-9 on 15 May 1963.

The Spacecraft

In the design of the Mercury spacecraft the goal was to keep it as simple and reliable as possible, with the minimum of new developments. The original concept for Mercury was just for a three-orbit vehicle, so when MA-9 was planned for a whole day in space, all the extra fuel, water, oxygen and electricity had to be incorporated into the spacecraft without adding too much weight. To keep it within the Atlas launch vehicle's weight lifting limits, 183 changes had to be made to the spacecraft. It was a small engineering miracle to keep Cooper's spacecraft within these limits.

For a Mercury flight the design engineers decided to use a drag-type re-entry spacecraft, an existing Intercontinental Ballistic Missile launch vehicle, retrorockets to bring the spacecraft back into the atmosphere, and a parachute to return to Earth. The Americans chose to make the astronaut more of a pilot in control of his spacecraft, able to override automatic systems, whereas the Soviets preferred automated systems to be in control with the cosmonaut playing a secondary role.

Maxime Faget, Head of the Performance Aerodynamics Branch in the Pilotless Aircraft Research Division, or PARD, was an ex-navy submariner and 38 years old in 1958. Faget preferred to experiment with real vehicles using information radioed back by telemetry, not studying models in a wind tunnel. With Benjamine Garland and James Buglia, he presented the NACA Report (reissued as NASA Technical Note D-1254) on 18 March 1958 with the uninspiring title of "Preliminary Studies of Manned Satellites, Wingless Configuration, Non Lifting", which became the basic working paper for Project Mercury to put America's first manned flights into space.[16]

A major consideration was the shape of the spacecraft. Originally, a spherical ballistic spacecraft was under consideration. Dr Robert Gilruth, Director of the Manned Spacecraft Center, said, "Harvey Allen of the Ames Aeronautical Laboratory was the first, to my recollection, to propose a blunt body for flying a man in space. In March 1958 Max Faget presented a paper that was to be a milestone in spacecraft design. His paper proposed a simple blunt body design that would re-enter the atmosphere without reaching heating rates or accelerations that would be dangerous to man. He showed that small retrorockets were adequate to initiate re-entry from orbit. He suggested the use of parachutes for final descent, and small attitude jets for controlling the capsule in orbit during retrofire and re-entry."[17]

In February 1959 wind tunnel tests of the spacecraft began. Scientists and engineers pored over schlieren photographs of shock waves, windstreams, boundary layers and vortices, and by the end of the year had tested over 70 different models. Spacecraft designers had to break away from the established streamlined concepts so popular at the time as the research by Harvey Allen indicated that his proposed blunt shape would dissipate 90% of the friction heat through the bow shock wave and the remaining 10% through the use of ablative coatings over a heatshield. Higher drag vehicles have less heating during re-entry than low drag vehicles. The drag of a streamlined vehicle is taken by the skin friction of the vehicle and all of the heat goes into the vehicle instead of the shock wave.

The Soviets chose a different system. For their re-entry vehicle they chose a sphere because Korolev, the Soviet's Chief Designer, felt that the sphere possessed an inherent dynamic stability as it entered the Earth's atmosphere. He felt the conical craft's tendency to pitch and yaw would require too elaborate an attitude control system. The Vostok spacecraft was aligned for re-entry by means of a solar sensor, which aimed the retrorockets to fire along the line of flight. After the rockets had fired the re-entry vehicle separated from the instrumentation section and became a simple sphere following a ballistic trajectory determined by the retrorockets and protected by an ablative heat shield that protected the entire spacecraft. By placing the centre of gravity behind and below the cosmonaut the proper orientation for ejection of the pilot was achieved. At 7,000 metres the hatch was explosively blown out to allow the cosmonauts and their seats to eject and begin the parachute descent from 4,000 metres to alight safely on land.

Senior Langley research engineer Hartley Soulé recalls that one of the main reasons the Americans rejected the spherical spacecraft was because the weight of the material required to completely shield the surface from re-entry heat would have made the spacecraft too heavy for their ICBM boosters at the time, whereas with the Soviets weight was less of a consideration with their more powerful boosters.

Following an industry wide competition, a formal contract for the research and development of 12 Mercury spacecraft was negotiated with McDonnell Aircraft Corporation on 5 February 1959. The first production spacecraft was delivered to Wallops Island for an abort test in early April 1960 and launched off the beach using its escape rocket to test the escape, parachute, landing, and recovery systems. This first trial was successful.[9]

Deke Slayton, one of the original Mercury astronauts, looked at the first Mercury capsule and found to his dismay there was only one little porthole down by the hatch. He said, "The hatch screwed on – there was no way to get it off. Ninety-to-one-hundred screws in the darned thing that had to be put on for the launch and taken off to get you out of there post launch.

"So the first three things we did: we got a window put up in front of the pilot so he could see and have some visibility to effectively control the machine. Then we had a control system in it that allowed us to control it. Then the third thing was we got a hatch that would blow off, so you could get out of the darned thing."

1959 Little Joe and Big Joe

Little Joe 1 began the Mercury flight test programme with an attempted launch from Wallops Island, Virginia, on 21 August 1959. Unfortunately, during the battery-charging phase during the countdown, the launch escape sequence was initiated by mistake and the spacecraft streaked off leaving the booster rocket behind on the pad. On its way back down the main parachute failed to break open due to a lack of electrical power and the spacecraft was destroyed on impact with the sea.

On 9 September 1959 an Atlas booster named Big Joe, topped by a boilerplate Mercury spacecraft model, roared off into the night sky above Cape Canaveral to check launch procedures and to see if the Mercury capsule worked. At the two minute mark the booster engine failed to separate, so the capsule never reached the planned height, but fell off at 106,680 metres to free fall back to Earth at over 24,000 kilometres per hour, 926 kilometres short of the target. Because of this the capsule was not recovered until after the press conference and some of the next day's headlines referred to the saga of the "lost" capsule. The heat shield temperature only reached 1,927°C, less than expected but enough to provide useful data on the shield, and the rest of the results proved enough to satisfy the Space Task Group officials that the capsule could align itself without an attitude control system, so the mission and spacecraft were declared a success.[16]

Project Mercury was under way.

America's First Astronauts

On 27 October 1958 a special committee on Life Sciences was established at Langley Research Center, one of the newly formed NASA establishments in Hampton, Virginia, to determine the qualifications for the pilots of America's first manned space flights. On 3 November the first military medical personnel reported for duty and began working on the human factors, crew selection, and crew training plans for the new manned space flight programme.[16]

Having determined that a man called an astronaut would fly into space, what sort of person should he be? There were no guidelines anywhere on Earth to define an astronaut. How big should he be? How strong should he be? What qualifications should he have?

During the autumn of 1958 NASA began looking at the type of man required for spaceflight. The initial thoughts were listed as:

1. Maximum age 40 years old, later to be lowered to 35 years. Young enough to be in their physical prime but old enough to have lost the rash impulse of youth.
2. Maximum height 180 cm.

Figure 2.3. The original seven Mercury astronauts. Standing, from the left: Alan B. Shepard Jr, Walter M. Schirra Jr and John H. Glenn. Sitting, from the left: Virgil I. "Gus" Grissom, Malcolm S. Carpenter, Donald K. "Deke" Slayton and L. Gordon Cooper.

3. Be in excellent physical condition with a maximum weight of 81.6 kilograms.

4. Hold a formal degree in Engineering or Physical Sciences, or their equivalent.

5. Be a graduate of a Test Pilot School.

6. Accumulated a flying time of 1,500 hours.

7. Be a qualified jet pilot.[9]

They wanted a pilot small enough to fit into the capsule cockpit, not be too heavy, mature enough for rapid but considered decisions, possess controlled aggressiveness with controlled passiveness for conditions such as long holds during a countdown, or when out of contact with ground stations. They were also looking for skilled military jet engineering test pilots, able to take over control of the spacecraft if the automatic systems failed.

Scott Crossfield, an X-15 test pilot and the first person to double the speed of sound, was quoted as saying, "Where else would you get a non-linear computer weighing only 160 lb (72.6 kilograms), having a billion binary decision elements that can be mass produced by unskilled labour?"

Navy psychologist Dr. Robert Voas, the Mercury Manned Flight Training Director, listed the characteristics he felt were required of an astronaut: Intelligence without genius, knowledge without inflexibility, a high degree of skill without overtraining, fear but not cowardice, bravery without foolhardiness, self confidence without egotism, physical fitness without being muscle-bound, a preference for participatory over spectator sports, frankness without blabber-mouthing, enjoyment of life without excess, humour without disproportion, and fast reflexes without panic in a crisis.[17]

"In short," one of the NASA officials was heard to say, "we are looking for a group of ordinary supermen."

A first culling of the personnel cards of qualified test pilots found 508 suitable candidates, who were reduced to 110 after consultation with their commanding officers. This was further reduced to 69 to be summoned to Washington, DC, where the candidates were informed about the Mercury Project and their expected role. At this point 37 pilots decided space was not real flying and returned to their units, which left 32 serious contenders to be subjected to a series of physical and psychological tests.

Only one of the 32 was found to have a medical problem serious enough to eliminate him, and 31 went on to Phase 4 of the tests at the Aeromedical Laboratory of the Wright Air Development Center. Eventually, on 2 April 1959 there were seven men left, seven men who defied every test the doctors and psychoanalysts could dream up. From the United States Marine Corps, Lieutenant Colonel John Glenn, Jr, received orders to report to the Space Task Group at Langley Field, on 1 May. There he found he was the senior astronaut in age and date of rank. From the Navy, Walter Schirra, Jr, and Alan Shepard, Jr, both Lieutenant Commanders and Lieutenant Malcolm Scott Carpenter. The Air Force assigned three Captains: Donald Slayton, Leroy Gordon Cooper, Jr, and Virgil Grissom. They became known as the "Original Seven" and attached the number 7 to all the names of their Mercury spacecraft to lionise this original team. On 9 April the first American astronauts were presented to the public in civilian clothes in Washington by Glennan, the Administrator of NASA.[9]

Later, Deke Slayton as the Director of Flight Crew Operations, developed a point system to grade his astronauts, which was broken into three categories: academic, pilot performance, and character/motivation with a maximum of ten points per category.[29]

The New Astronauts Suffer Gruelling Tests

Like the Soviets, the Americans subjected their long-suffering astronauts to torturing tests to prepare them for what they believed was in store for them. Some of the tests at the time indicated that a man could be shaken to death by sympathetic vibrations induced through various harmonics upon certain organs. They walked and ran on endless treadmills, sat with their feet in iced water, rode stationary bicycles while the brakes was slowly tightened, swam underwater in tanks of water, blew up balloons until exhausted, baked for two hours in a room heated to 54°C and endured three hours in a soundproof, pitch dark chamber.

The 'g' Factor

Little was known about the effect of extreme forces on the human body before space travel so it became quite an obsession with the early space fraternity. The forces of acceleration and deceleration are measured in 'g's, with 1 g for normal gravity on the Earth's surface. From '0' g, or weightlessness, to – well, what could a human stand? What forces would they have to withstand to survive flights into space?

During World War II many fighter pilots had experienced momentary pain and blurred vision during "redouts" when the blood pooled in the head and eyes during outside loops, or "blackouts" when the heart could not pump enough blood to the brain to cope with the forces draining the blood out of the head when pulling out of dives. Physician Harald von Beckh at the Aeromedical Field Laboratory in New Mexico was concerned how a space traveller would cope with the high deceleration forces of atmospheric re-entry after days of weightlessness. There was no way of experimenting with days of weightlessness on the Earth.

Although there had been research for around twenty five years, detailed knowledge about the effect of these g-forces in the early fifties was still quite hazy. Having exhausted the aircraft as a tool for studying g loads, the aeromedical scientists turned to two devices on the ground – the rocket-powered impact sled for studying the immediate onset of g-loads, and the centrifuge, where a slower build up of g-forces could be simulated.

In May 1958 Captain Eli Beeding experienced 83 g for 0.04 of a second at Holloman Aerospace Medical Center. After recovering from shock and various minor injuries, Beeding figured that 83 g represented the absolute limit of human tolerance. From these experiments it was determined that the maximum speed an aft-facing, contour couched astronaut could survive a spacecraft impacting the ground would be 143 kilometres per hour, providing the vehicle remained intact.

After trying various devices including total immersion in water, in 1958 NACA engineers led by Maxime Faget designed a fibreglass contoured couch to maximise the ability of a pilot to survive the acceleration and deceleration forces expected. Through April and May of 1959 McDonnell engineers test dropped four Yorkshire pigs in contour couches to experiment with crushable aluminium honey-comb impact systems for landing the spacecraft on land. The pigs suffered acceleration/deceleration peaks from 38 to 58 g before minor internal injuries were noticed. When the pigs climbed out and walked off after a drop the scientists and astronauts hoped that man would be able to do the same. Pigs were no longer used as test subjects when it was found they could not endure long periods lying on their backs.

On 30 July 1959 Lieutenant Carter Collins, using a grunting technique to avoid blackouts and chest pains, experienced loads of up to 20.7 g applied transversely for a period of up to six seconds. The contoured couch used by Collins, with adjustable body angles, was the system selected for the Mercury spacecraft.

So for this new breed of humans called Mercury astronauts there were endless rides in a human centrifuge, spinning them all ways to crush their bodies with forces, and pressure suit runs to a simulated 19,812 metres altitude. After training, the astronauts found that they could perform certain tasks up to 14 g providing they were prepared for it and properly supported, although 8 g was the maximum expected. In a letter to a pilot friend, John Glenn said, "We spent a couple of weeks this fall (1959) doing additional centrifuge work up at Johnsville. This was some programme since we were running it in a lay-down position similar to that which we will use in the capsule later on and we got up as high as 16 g. That's a batch in any attitude, lay-down or not.

"With the angles we were using, we found that even lying down at 16 g it took just about every bit of strength and technique you could muster to retain consciousness. Some other stuff we did up there involved what we called tumble runs, or going from a plus g in two seconds to minus g and the most we did was going from a plus 9 g to a minus 9 g."

This was done by flipping the gondola containing the astronaut over in the middle of a run.[9]

Glenn said the large centrifuge, which they called the "Wheel", was the most spectacular of all the stress making machines used in their training, and he called this exercise "Eyeballs Out, Eyeballs In".[20] After each test the astronauts were checked for how long it took them to recover from each stress test.

The g-factor as a problem melted away with experience and careful rocket design. The maximum forces experienced by the Apollo crews was up to 4 g and with the shuttle flights the maximum forces experienced by the crew does not normally exceed 2 g.

Other Tests

Continuous psychiatric interviews, the necessity of living with two psychologists throughout the week, and extensive self examination through a battery of 13 psychological tests for personality and motivation, and another dozen different tests on intellectual functions and special aptitudes were all part of the week.

Then there were the searching psychological tests, to measure their maturity, mental alertness, motivation, and find out what kind of person each candidate was. They had to answer questions such as "Who am I?" with twenty different responses, and reply to 600 questions on their personalities. In one test a blank sheet of paper was put in front of them and they were asked to describe what they could see in it. Pete Conrad retaliated by turning it around and pushed it back saying, "It's upside down."

They were trained by experts for survival in deserts, jungles, and the open sea. They flew special flight paths in aircraft to experience weightlessness, and were plucked out of the sea by helicopters. They studied astronautics, aviation, biology, astronomy, meteorology, and astrophysics. They had to be familiar with all the systems of the Mercury capsule, launch vehicle, navigation, and tracking facilities, and finally they had to specialise in the design and operation of one of the spacecraft systems. As well as all the above they had to be cool in a life-threatening emergency, which was put to the test many times as they moved down the track to the Moon. Gus Grissom felt that the training, particularly for Gemini, was like working a 48-hour day in a 14-day week.

American Space Tracking Stations Set Up Around the World

It is interesting that nearly all of the early fiction and theoretical proposals ignored the requirement for the crews to keep in touch with their base. Their heroes were sent off into the wide black yonder to return with their exciting adventures much as the sailing

Figure 2.4. On the white sands of Muchea. The Telemetry & Control Building with the Mercury Acquisition Aid and Timing antennas.

ships of yesteryear. It seems that the minds of the time were conditioned to think of the spacecraft as ships – sent away and expected to fend for themselves. With the complex spacecraft and technology of the Apollo era, it was mandatory to keep in constant contact with the skilled engineers and procedural experts on the ground to carry out their voyages, particularly during emergencies, which is why the specialised tracking networks were built. This concept was to prove vital in the manned missions to come, with Apollo 13 the best example. Also, thanks to this system, the people of Earth had the luxury of witnessing the adventures of the Apollo missions as they happened. Later, when the Shuttle missions began, these many ground stations spread around the world with only brief glimpses of the passing spacecraft were replaced by more efficient communications satellites in space able to keep in touch with the spacecraft at all times, with only minimal ground stations. In the Mercury days there weren't the satellites in space, and the communications from the overseas stations back to the United States, using radio and undersea cables, was vital. A world-wide network of specially designed and built tracking stations were planned for keeping in constant touch with the increasing number of spacecraft and satellites being pushed into space.

The global and planetary nature of the missions meant that the mission control centres needed to keep in contact with their spacecraft, especially at high activity periods such as landing on the Moon and planetary encounters. The only way to do this was to have tracking stations spaced at least about 120° around the world. With a logical location of a station in America, the other two would fall roughly at the longitudes of Spain and Australia, and these were selected as being the most suitable sites. With the manned Earth orbit missions, tracking stations were needed at frequent intervals along the flight path, and Mercury had up to 17 stations, including two ships, one in the Indian Ocean and one in the Atlantic. At about the same time as the first tracking stations were being organised, teams of flight controllers were being selected to run them. Project Mercury required quick entry into the foreign countries for purposes of site selection and development while formal negotiations were under way as the original schedule called for a worldwide operational tracking network by early 1960, but luckily flight schedules slipped and a deadline of 1 June 1961 was eventually met, which turned out to be well ahead of the first orbital mission on 20 February 1962.

1960 The Australian Mercury Tracking Stations Chosen

Muchea, just north of Perth in Western Australia, was a key station in the Mercury network, situated roughly at the highest point of the spacecrafts' orbital path, and was a Command station, which meant that it could send instructions to the spacecraft. The next Command station wasn't until Hawaii.

At the time the "town" of Muchea only consisted of a typical country general store run by an elderly lady, Mrs Blanche Peters. She also manned a three-line telephone switchboard, her window looking out on the paddock alongside at stacks of 200 litre fuel drums.

"My switchboard is essentially a local service. If anybody gets sick in the district and they want a doctor, I'll have no choice but to cut them off – Australians or Americans alike. I must tend to my own folk," Mrs Peters, actually an avid follower of space activities, told a reporter.

Kevyn Westbrook, the Ground Communications Coordinator, remembers, "There was no such thing as microwave communication in those days, it was all copper wire land lines. We used to have a lot of outages because people in their cars were always running into the poles, and the lines were getting tangled up with galahs, the native wild bird. The galahs would be feeding on the ground when a car would go past, and they would all rise up in a flock and get tangled up in the wires, and once they got all wrapped around that was the end of the circuit as we knew it.

"We needed a 'spacecraft' to practice our tracking, so we fitted out a RAAF DC3 aircraft like a spacecraft so we could practice our procedures. After a while it became a little boring as we'd sit there while the plane would drone overhead and we would go through the standard mission routines.

Figure 2.5. The Verlort radar at Muchea. This military "S" Band radar was housed in trailers and had a range of 1,600 kilometres.

dreamed of, the first to blaze man's trail to the stars – name a task more complex than the one I am facing. I am not responsible to one man, or only to a score of people, or merely to all my colleagues. I am responsible to all of the Soviet people, to all mankind, to its present and its future. And if, in the face of it all, I am still ready for this mission, it is only because I am a Communist, inspired by the examples of unsurpassed heroism of my countrymen – the Soviet people."

Figure 2.11. "See you all soon."

At 7:00 am Gagarin went up to the Vostok spacecraft on top of the 300 tonne modified A1 ICBM rocket, 30 metres from the ground. He turned around, raised his arms to salute the small crowd gathered below and called out, "See you soon." Then he turned and entered the spacecraft, assisted by Oleg Ivanovsky, one of Vostok's leading designers, and by Mark Gallai, the Chief Test Pilot.

Titov watched the hatch close on Gagarin and for a few moments tried to imagine what was going on inside, how he was feeling. He wanted to help but could do nothing more, just watch with the rest of the small crowd.

During the one and a half hours of testing, almost as long the flight itself, Korolev spotted an electrical contact on an attachment ring was not making, so the access hatch had to be removed and replaced before all was ready, and the final count down began.

On top of the rocket, sitting in his spacecraft he had named *Swallow*, Yuri Gagarin was at the very tip of a pyramid of every space traveller that will ever leave the Earth until the end of human existence. In the whole history of the Earth he was the first human to rise above the atmosphere and enter space.

Gagarin's launch seemed to be more relaxed than NASA's militaristic drill. For instance the Russians had no public address system and did not use the suspense-generating countdown to the moment of lift-off adopted by the Americans. At twenty minutes to launch Gagarin queried ground control, "What do the doctors say, is my heart still beating?"

Ground Control responded, "Your pulse is 64, breathing 24. Everything is normal."

"Roger, that means I am still alive!" quipped Gagarin

Control called Gagarin on the intercom: "Yuri, are you getting bored out there?"

"If there was some music I could stand it a little better." Korolev immediately organised some tapes to play music for Gagarin while he waited.

At 9:07 am, Moscow time, the Range Controller in the Command Post pushed the button and the engines fired.

Gagarin heard "Idle run" in his earphones, then "Ignition".

Tensing his muscles, Gagarin yelled, "*Poyekhali!*" (Let's go). He found he could not pick the exact moment of lift-off with all the vibrations and rumblings going on under his

Figure 2.12. *"Poyekhali!"*

seat: "I heard a shrill whistle and a mounting roar. The giant ship shuddered, and slowly, very slowly, lifted from the launching pad. The acceleration of g-forces began to increase. I felt that some uncompromising force was riveting me to my seat. I could hardly move my hands or feet."

The Earth around the base of the rocket exploded; grass and bushes hundreds of metres away were blown flat by the exhaust gases as they pushed their way out of the thundering motors with tremendous force, making a white, energy packed cloud that was awesome to watch. For a fraction of a second the rocket seemed to hang in the air – then it was away, gathering speed smoothly and steadily, pushing itself out of the smoke and fumes, streaking towards the sky. In seconds Vostok I had tilted slightly to find its course and to the watchers on the ground the spaceship became only a dwindling dot in a vast expanse of blue.

Gagarin: "Here we go. The noise is faint in here. Everything is going according to plan, I feel fine."

Korolev: "We wish you a good flight, everything's all right here."

Gagarin: "I'll see you soon, good friends."

Inside the capsule's confined cabin, Gagarin was being pressed back into his couch with the fierce acceleration. At 119 seconds the big A-1 booster dropped off and Korolev called, "How do you feel?" There was no answer, Gagarin was finding it hard to talk into the microphone with the g-forces, though he felt it was no worse than doing a tight turn in a MIG fighter. The people on the ground strained to hear any sound from the spacecraft, but the earphones and speakers just hissed back at them.

"Answer me," ordered Korolev sternly; he regarded Gagarin as a son, but there was still no response. Korolev was becoming worried – there were so many unknowns and potential equipment failures to consider. In the spacecraft Gagarin was being flung forward in his straps as the booster finished firing and dropped off, before the second stage took over the job of shoving the vehicle upwards. Three minutes after lift-off the streamlined shield blew off the top of the spacecraft and Gagarin could see the darkening blue of space through the porthole. Vostok 1 went into orbit 11 minutes 16 seconds after leaving the ground. Instead of silence Gagarin became aware of the constant noise of the motors of the fans and pumps, with the harsh hiss of the radio receiver in the earphones. At first he could only see the blackness of space through the porthole.

Gagarin leaned forward looking for a view of the Earth. *Swallow* was slowly rotating and as he peered out of the porthole the intense black of space was replaced by a translucent blue – the first time a human had seen the Earth from space. "The surface of the Earth loomed in the distance through the portholes. Just then the Vostok was flying over a broad Siberian river. I could clearly see sunlit, taiga-covered islands and banks." He became very aware of the curved horizon and found he could clearly see mountain ranges, large rivers, big forests, islands, and the coastlines of land.

"Weightlessness has begun. I can see the Earth in a haze. I can see clouds. Feeling fine. How beautiful …," the words crackled out of the speakers down on the ground and everyone breathed again – he was all right after all. Excitement grew with the tension as they realised Gagarin was speaking from space. They had actually succeeded to put a man into orbit around the Earth!

"I found myself in the state of weightlessness, about which I had read in books by Tsiolkovsky when I was a boy. At first I felt uncomfortable, but I soon got used to it. I suddenly found I could do things much more easily than before. And it seemed as though my hands and legs and my whole body did not belong to me. They did not weigh anything." Gagarin felt the pressure on his back against the couch had eased, so loosened the seat belts, and immediately floated off the seat. There were no unexpected feelings, no strange sensations, just pleasant relaxed freedom. As he reeled off instrument readings to ground control he noticed the velocity dial was showing a steady 27,358 kilometres per hour, and the cross hair of the global position indicator in front of him showed him steadily passing over the Earth's surface, over Siberia, over the Pacific Ocean.

The Americans tracked the flight around the world, one group gathered around tracking equipment on a tiny atoll called Tern Island in the French Frigate Shoals, 724 kilometres northwest of Honolulu. Thirty minutes after he was launched, just after the Sun had set on the Earth's surface, but was still shining on the spacecraft, it was clearly visible to the watchers below. An operator followed the bright spot of light through a set of ×25 power binoculars, and they picked up and recorded the radio signals by slaving to the operator manning the optical tracker. They were able to see the cosmonaut's heart beat on the signal coming down from the spacecraft. There was a tracking signal transmitted at 19.995 MHz, while communications were on 9.019 MHz, 20.006 MHz, and 143.625 MHz.

Then *Swallow* plunged into darkness, and during the few minutes of night, Gagarin tapped out VOSTOK on a telegraph key, and heard ground control verify reception in his headphones. He had hoped to see the Moon, to see how it looked from space, but it never came within his field of view. "Next time," he thought to himself.

Gagarin sang happily as he went about the tasks set down by the flight plan, but was interrupted by the ground controller who commented, "When you're through singing, we've got a professional," and the strains of Moscow Nights flooded his earphones.

While most of the American continent was sleeping through the history making flight, Vostok passed over Cape Horn at the bottom of South America. In the night sky above Gagarin could see the stars, more than he had ever seen before, and he noticed they were steady, not twinkling as seen from the ground. Gagarin was instructed to send a code with his telegraph key every few minutes while out of voice communications, and once the signal reported a malfunction, and for a moment Korolev feared trouble, but it corrected itself. "It's seconds like that which shorten a designer's life," he remarked with relief.

Approaching his second dawn of the day, the first human to experience two dawns in one day, Gagarin watched as the first rays spread through the hazy atmosphere, to lead to a fiery orange glow on the horizon, changing to all the colours of the rainbow. "It reminded me of the canvases of the artist Nikolai Rörich," he wrote. Gagarin was pleased to see the cabin temperature was still around a comfortable 20°C; the pressure was holding, and the humidity steady at 65%. He wrote his name and VOSTOK on a pad strapped to his knee, before reaching for the first meal in space of a meat-like jelly in a fountain-pen-sized tube, prepared to a recipe by the Academy of Medical Sciences. He had a drink of water from a special supply system. By the time he had finished squeezing some of the salty paste into his mouth it was 10:15 am and they were over Africa, and a red light glowed to warn him that they would be descending in the next ten minutes. It had taken three years for Magellan's ships to sail around the world, but Gagarin was round in an hour and a half! He noticed that the Sun shone ten times more strongly in space than on Earth. While busy writing down all his impressions, Gagarin reported to ground control everything was in order and the spacecraft was ready for an automatic re-entry.

Gagarin strapped himself down, lowered the seat into the fully reclining position and braced himself for the unknown forces of re-entering the atmosphere. First the instrument section should drop off leaving the spherical re-entry capsule with its cosmonaut to plunge into the atmosphere like a meteor. Now he was about to be the first human being to return to Earth. "What would this last, final stage of the flight be like? Would all the systems operate normally? Was some unforeseen danger lying in wait for me?" he thought.

At 1 hour 18 minutes from launch the retrorockets fired with a roar, and Gagarin sensed the cabin atmosphere around him change, then felt the pressure of his weight returning. The separation of the capsule from the instrument module occurred on time but a cable connecting them did not disengage initially and the two units began to spin around each other, threatening to put the capsule out of alignment for re-entry. On the instrument panel the indicator lights went out then lit up again. Gagarin now felt something was wrong, there had been no proper separation. He was spinning about in three axes and he could feel the capsule heating up. A crackling sound told him the capsule or the heat shield was expanding with the heat. As expected, all radio contact with the ground was lost due to the ionisation of the capsule.

When the heat of re-entry vaporised the cable, the capsule and instrument module finally separated to head off in different directions, the instrument module to burn up, and Gagarin to re-enter the atmosphere behind the protection of his heat shield. He gazed in awe at the fiery spectacle of re-entry through the portholes. Long tongues of white flames streaked furiously past and stretched out of sight behind. This was the moment nobody knew what would happen to a human trapped inside the spacecraft.

"Through the portholes I saw the crimson glow of the flames that raged around the ship. I was a ball of fire plunging downwards. Weightlessness had long ceased, and the mounting overload pinned me to my seat. They kept increasing and were much greater than the stresses I had experienced at the take-off," Gagarin noted. The pressure built up to almost painful proportions, every nerve and sinew was hammered by vibration as the capsule spun, at one point so fast the instrument panel blurred and everything went gray. After one and a half minutes the noise and pressure eased as an orange light on the panel in front of him drew his attention to prepare for landing, and radio communications with the ground were restored. A pale blue sky through the porthole told Gagarin he was back in the Earth's atmosphere. He reached for the manual ejection cord in case the automatic system failed.

Figure 2.13. Gagarin's Vostok 1 lying on the stubble of a Russian field. "In general, all my feelings can be expressed in one word: joy," was how he summarised his flight.

At a height of 7,010 metres, the main hatch was automatically blown off and Gagarin suddenly felt very vulnerable as the tiny cabin was opened to the elements.

"Above target, para-brakes," announced ground control, and at 4,000 metres braking parachutes broke out to slow both the capsule and cosmonaut until the main parachute's canopy spread out at 2,500 metres. The Russian ground controllers advised him not to eject too early and to wait for automatic ejection, but suspecting there could be trouble ahead with the spinning capsule, Gagarin chose to eject manually before the scheduled time, and shot out of the spacecraft. He was soon floating softly down through the familiar blue of the Earth's atmosphere. He was so happy and relieved to see the patchwork quilt of the farmland around the district of Saratov appear beneath his feet that he burst into his favourite song, "The country hears, The country knows…"

He later reminded us, "I returned from outer space to exactly the same spot where I first learned to fly an aircraft. How much time had passed since then? Not more than six years."

Like all Vostok cosmonauts, Gagarin had to eject from his capsule in his couch before landing as the impact was expected to be too much for a human. Initially,

Soviet officials had announced that Gagarin had landed inside his capsule to satisfy what Soviet officials believed were the international rules to qualify Gagarin as the first space-man to return to Earth.

At 10:55 am Anna Takhtarova, the wife of a local forester, was weeding potatoes in the fields, southwest of the little town of Engels. Her grand daughter, Rita, was playing nearby. They looked up to see a strange object appear in the sky. As they watched one said to the other, "Parachutists!" and they began running towards the cluster of chutes as they skimmed over some trees and hit the Earth in the middle of a stubbly field. They stared in fear and amazement at the orange-suited cosmonaut and his parachute.

"Hello! I'm a friend, comrades," he grinned.

At the sound of the Russian voice, the young girl giggled, and her grandmother asked, "Have you really come from outer space?"

Gagarin replied, "Just imagine it, I certainly have."

She later told reporters, "I was a bit scared. The man's clothes were strange and he appeared out of the blue. Then I saw him smile – and his smile was so good that I forgot I was scared."

Yakov Lysenko was driving his tractor in the fields and heard a bang in the sky above. Looking up he saw Gagarin's parachutes open and thought it was a pilot from a plane, and ran to the village of Smelkovka to raise the alarm. Collecting a group of friends they went to meet the pilot. Gagarin approached them and shook their hands saying, "I am the first spaceman in the world – Yuri Alexeyevich Gagarin."

Next, officials from a nearby military camp, led by General Stuchenko, raced up in a car, to whisk Gagarin off and the first manned space flight in the history of the world ended. Gagarin went to a maximum height of 327 kilometres with a flight lasting 108 minutes in Earth orbit.

The whole of the Soviet Union rejoiced. School children, factory workers, public trans-port – everybody stopped to listen to their radios, or watch television. Premier Nikita Khrushchev called Gagarin from his holiday home on the Black Sea, "By your feat you have made yourself immortal. Let the whole world see what our country is capable of, what our great people and our Soviet science can do."

When asked about his feelings on returning to Earth, Gagarin replied, "It is difficult to say in words all the feelings that took hold of me when I stepped on our Soviet land. First of all I was glad because I had successfully fulfilled my task. In general, all my feelings can be expressed in one word: joy."

Yuri Gagarin was so confident about the trip he never left a will or wrote a farewell letter to the wife he loved so dearly, and told his parents he was only going away on a business trip.

After the successful flight, Alexander Topchiev, vice president of the USSR Academy of Sciences, said, "The battle to put a Soviet astronaut on the Moon is half won. It will be done."

1961 America's Reaction to Gagarin's Flight

In America, early on the day of the launch, a reporter rang Lt Col. "Shorty" John Powers, press officer for the seven Mercury astronauts, and asked, "Did you know the Russians have put a man in orbit and recovered him?"

"It's three am in the morning, you jerk," replied Powers. "If you're wanting something from us, the answer is we are all asleep."

The House of Representatives member Thomas Pelly reporting this episode on the floor in Washington, said, "I suggest that we went to sleep in 1945 and slumbered until the Eisenhower Administration got us under way in 1953. By that time the Russians had established their tremendous lead in payload rocketry and today we have reaped the harvest of yesterday's neglects."

President Kennedy bravely admitted to his public, "We are behind. The news will be worse before it is better, and it will be some time before we catch up."

By now *Pravda* of 13 April 1961 could proudly announce to the world that: "The first person to penetrate outer space was a Soviet man on the 12th of April, 1961. The Soviet Union was the first to fire a ICBM, the Soviets were the first to send a probe to the Moon, the Soviets created the first artificial satellite of the sun, Soviet orbital spaceships with living creatures on board carried out flight in space for the first time."

When Kennedy's administration took over the reins early in 1961, Jerome Wiesner, Kennedy's science adviser, was chairman of a committee that was critical of NASA's record to date, and doubtful of its future. Kennedy was advised not to push the manned space flight business, to which he agreed at the time, but the fledgling NASA, led by their fiery new Administrator, James Webb, was urging Kennedy to approve the Apollo Project. Looking back, Wiesner remembered that Kennedy wasn't terribly excited about doing it at the time – something not generally understood. Webb, a fast talking, dynamic manager, was everywhere at once and seemed to know everyone who mattered. He won the support of industry, brought academia on side, sold the Moon project to Congress, and infused his listeners with eager enthusiasm for space to such an extent one Congressman remarked, "Listening to Jim Webb is like taking a drink from a fire hydrant."

Gagarin's flight infected the Americans with space fever, and with Webb's enthusiasm, stoked the fire of Congress to the point of offering NASA a blank cheque. In fact Dr Robert Seamans, NASA's Associate Administrator, said he was hard put to restrain Congress from forcing more money on NASA than it could handle!! A comment he was to regret later.

1961 **NASA's Mission Control Center to Move to Houston**

On 19 September 1961 NASA Administrator James Webb announced

What is the difference between "orbit" and "revolution"?

You will often see these two terms used in mission descriptions. Do they mean the same thing? Not really, although they are related. The NASA Manned Space Flight Network used the following definitions:

An **orbit** is the path of a spacecraft around a celestial body, say the Earth, from a point in space back to that same point in space, irrespective of the position of the planet below.

A **revolution** is a complete circle around a celestial body starting and ending above the same point on the planet's surface. A revolution has not been completed until the spacecraft has travelled the extra distance that the starting point has travelled during the period of the orbit as the planet rotates. In the case of the long Gemini VII mission lasting 14 days, the spacecraft actually completed 220 orbits, but only 206 revolutions.[23]

In Figure 2.14, one orbit is from point A, around the Earth, and back to point A. Since the longitude of Cape Canaveral has rotated about 23° to the East (point B), one revolution has not been considered completed until the spacecraft has travelled the additional distance, and is again over the longitude of Cape Canaveral.

Refer Figure 2.14 overleaf

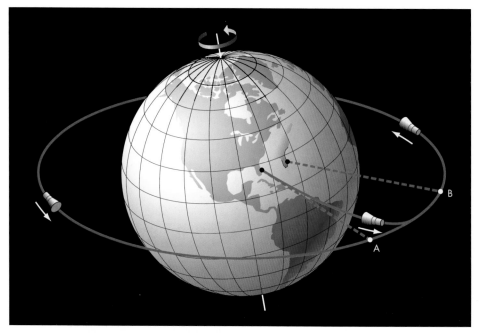

Figure 2.14. The difference between an "orbit" and a "revolution".

that a new, bigger Mission Control Center would be built in Texas on land donated by Rice University, Houston. The Space Task Group would move from Langley Field and the Operational Center would move from Cape Canaveral, to Houston, which was to become synonymous with the Gemini and Apollo programmes, and later, as the Johnson Space Center with the Space Shuttle.[16]

1961 Mercury-Redstone 3: America's First Manned Flight into Space

The Redstone rocket sat on the launch pad like a slim digit pointing to the heavens – and the future. Jammed tightly into the spacecraft perched on the very top was US Navy Commander Alan B. Shepard, Jr, lying on his back waiting to be the first American into space. His only view of the outside world was through a retractable periscope. Two tiny portholes were covered by a shield that would blast off with the escape rocket.

Shepard felt the choice of the name *Freedom 7* for his spacecraft was self-explanatory. The "7" was attached to the name because it was the seventh spacecraft off the assembly line, but by Gus Grissom's flight the seven became a symbol of the first seven Mercury astronauts and was attached to all the Mercury spacecraft names. *Freedom 7* would have

Mission Data: Mercury-Redstone 3	
Date:	5 May 1961
Craft name:	*FREEDOM 7*
Personnel:	Alan Shepard
Duration:	15 minutes 22 seconds
Features:	Sub-orbital
Max altitude:	187.5 km
Max range:	486 km
Max velocity:	8336.2 kph
Spacecraft weight:	596 kg

Figure 2.21. "Astronaut to Control – it's turned out nice again, hasn't it? ....Control? ...CONTROL!!!"

Glenn strode past the cluster of well-wishers with a wave and was driven out to Pad 14, accompanied by weatherman Ken Nagler to brief him on the current weather picture. "At the pad we opened the blinds on the van windows, and I could see the gantry. It was a beautiful sight. It was still dark, but the big arc lights shone white on the Atlas booster. It looked like something out of another world."

On the other side of the world at Muchea, the station air-conditioning was struggling to cope with sweltering temperatures of 40°C. After so many false starts, there was nothing to indicate that this was going to be the day. As had been their custom, the American flight control team of Astronaut Capcom Gordon Cooper, Assistant Capcom Stanley Faber, and Flight Surgeon Dr. E. Beckman, called in for a while to check on the status of the mission, there being no such luxuries as mobile phones in those days.

The interminable holds continued to interrupt the countdown, but finally at 6:00 am Glenn arrived at the gantry. Coming up to T-minus-120-minutes, Glenn rode the elevator to the eleventh deck and met Scott Carpenter, the back-up pilot. "Scott was standing off to one side, out of the way, and I walked over to him for a moment. I know something about being a backup pilot, having done the job twice myself. It is hard work, and the personal satisfaction is limited to helping someone else. Scott had pitched in from the beginning as if he were preparing for his own flight, not someone else's, and I appreciated this very much. Scott was my alter ego."[20]

Glenn climbed into the spacecraft and settled down to an hour of checking procedures before the hatch was closed and sealed. But he had an encore. One of the seventy hatch bolts broke while being tightened down, and it was forty minutes before the hatch was finally secure. Lying there in his suit, moulded into the contoured couch, Glenn could feel the tall pencil-like rocket squirming and vibrating under him as the metal skin flexed, a vertical column of liquid fuel only held in place by a thin metal skin which rippled and shuddered to every movement, even to gusts of wind. Glenn found that he could start the whole structure wobbling a bit just by jerking around in the couch.

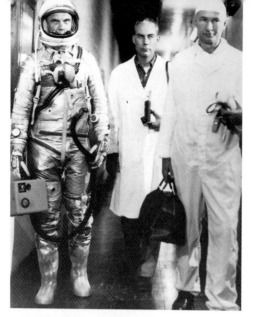

Like Shepard and Grissom, he found a relief from the tedious constant checking by peering through the periscope at the limited view. He could see the Sun sparkling off the cheerful blue water of the Atlantic one way and swivelling the periscope around could just make out the thousands of people gathered around the beaches and swamps to watch and be part of this moment of history. Some of them had been camped there in temporary communities for nearly a month. An estimated 60 million people watched the launch on live television.

In Mission Control Alan Shepard was sitting at the Capcom's console in front of Chris Kraft, talking to Glenn, "I could hear his voice clearly as he read off his meters and gauges to me. He sounded exhilarated and in full command of the situation. When I knew Chris was looking at me I stuck out my left hand and raised it, thumb up. We were on our way."

At 35 seconds to go Glenn watched the umbilical cord which provided the power and cooling drop off, the last connection with the ground equipment. All the tracking stations around the world were "Green", with everything ready to support Glenn.

"God speed, John Glenn," said Scott Carpenter, and the engines fired. Glenn

Figure 2.22 (top). After 82 days of delays, John Glenn wonders, "Will we make it this time?" as he heads for the spacecraft, escorted by Dr Bill Douglas and Suit Technician Joe Schmitt...

Figure 2.23 (middle). ... to become a "man in a can" as he climbs into *Friendship 7* on top of the Mercury–Atlas rocket...

Figure 2.24 (bottom). ... that sent him into space to become the first American to orbit the Earth successfully.

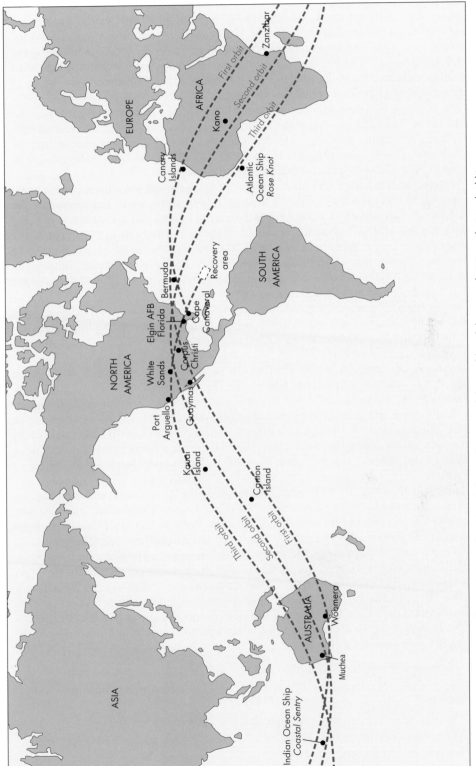

Figure 2.25. John Glenn's three orbits around the world and the ground stations that tracked him.

recalled, "The spacecraft shook, not violently but very solidly. When the Atlas was released there was an immediate gentle surge that let you know you were on your way. The roll to the correct azimuth was noticeable after lift-off. I had preset the little window mirror to watch the ground. I glanced up after lift-off and could see the horizon turning. Some vibration occurred immediately after lift-off. It smoothed out after about 10 to 15 seconds of flight, but never completely stopped."

The ride was not slow and sluggish as he had expected from seeing previous Atlas rocket launches. At 20 seconds into the flight Glenn checked through the instruments and found everything working perfectly. At 10,668 metres, 45 seconds after launch, the rocket entered a dangerous period of vibration. On one of the trial launches with an empty capsule, an Atlas rocket had blown up at this point.

Glenn said, "Just before the end of powered flight there was one experience I was not expecting. At this time the fuel and 'lox' tanks were getting empty and apparently the Atlas becomes considerably more flexible than when filled. I had the sensation of being out on the end of a springboard and could feel oscillating motions as if the nose of the launch vehicle was swaying back and forth." He strained against forces that built up to about 6 g before he was weightless as the vehicle curved over to enter orbit. Although he was quite comfortable Glenn wanted to see more than the window allowed. He quipped later, "I guess I'd like to be in a glass capsule."

As *Friendship 7* sped over the Indian Ocean, Glenn saw his first sunset in space. After the Sun had gone, the Earth went dark, looking down it was like gazing into a dark pit, but then an almost full Moon appeared and brightened up the view, the clouds showing up clearly.

Approaching Australia, it was 11:37 pm in Perth as Glenn was busy studying the stars, but he turned around and looked for the lights of Perth. Glenn recounted, "Through the window, I could see several great patches of brightness down below. Gordon Cooper who was on duty as Capcom at the tracking station at Muchea had alerted me to look off to the right. He knew that the citizens of Perth and several other cities and towns along the west coast had turned on all the lights they had as a greeting, and when I spotted them I asked him to thank everyone for being so thoughtful."[20]

Isolated in their van, Don Blackman was manning the Verlort radar with Ken Lee and Norman Hurrel. Blackman remembers, "I was controlling the antenna and Ken was watching the receiver and the transmitter. We knew the signal was due to appear on the screen at any moment, and I certainly remember the tension, and then the great relief when we saw the blip appear on the screen and we knew we had locked onto the capsule. From then on we just tracked the capsule until it disappeared off the screen. We didn't know what was going on in the rest of the station at all."

The first Australian to talk to a space traveller was Jerry O'Connor, the Communications Technician manning the Air/Ground console at Muchea. In the first flights it was his job to establish initial contact with the capsule, and hand over to the Capsule Communicator, or Capcom.

O'Connor: "Friendship Seven, Muchea Com Tech. We read you. Would you–"

Glenn: "Hullo, Muchea Com Tech. This is *Friendship 7*, reading you loud and clear. How me?"

Cooper: "Roger, *Friendship 7*. Muchea Cap Com. How me? Over."

Glenn: "Roger. How are you doing Gordo? We're doing real fine up here. Everything is going very well. Over."

Cooper: "John, you sound good."

The last entry in a standard post mission form asked, "Was there any unusual activity during this period? " With a grin, Glenn filled in, "No. Just a normal day in space."

He earned $245 in pay for his 4 hours in flight.

In Western Australia astronaut Gordon Cooper, Capcom at Muchea, received a message from Cape Canaveral during a station debriefing, and was able to tell the staff, "Muchea led all the other stations in the network on all three passes on radar coverage and accurate sightings. It got more coverage than any other site, and got the maximum amount of data."

Led by Cooper and Station Director Lewis Wainwright, the staff celebrated the splashdown with a party at a suburban hotel. Mayor Howard was there to help them celebrate.

It was announced that 26 February was "John Glenn Day" in Washington, with a White House reception, parade, and an address to Congress. President Kennedy tactfully escorted Glenn to the car outside the White House, and left him to face the limelight. On 1 March, another "John Glenn Day" saw four million cheering, yelling New Yorkers turn out to welcome and glorify Glenn and the astronauts in a deluge of shredded telephone books and paper. The astronauts were now national and international heroes.

Glenn found weightlessness was a pleasant experience. He could move his head around vigorously with no unpleasant effects. Exercise produced the same result as on Earth – he became tired. He had no difficulty adapting to zero gravity, "I found myself unconsciously taking advantage of the weightless condition, as when I would leave a camera or some other object floating in space while I attended to other matters. I thought later about how I had done this as naturally as if I were laying the camera on a table.

"Even where automatic systems are still necessary, mission reliability is tremendously increased by having the man as a backup. This mission would almost certainly not have completed its three orbits, and might not have come back at all, if a man had not been aboard," Glenn said, defending NASA's programme to send pilots into space.

Figure 2.30. The rewards of success: John Glenn with President Kennedy and Major-General Leighton Davis, Commander of the Canaveral missile launching complex, driving through Cocoa Beach in Florida.

The erroneous signal inferring the heat shield was loose to ground control turned out to be a limit switch with a bent and loose shaft, which could be operated to send an alarm signal without appreciably displacing the sensing shaft.[22]

In October 1998, thirty six years after this flight, John Glenn, aged 77 years, went from the oldest Mercury astronaut to the oldest astronaut into space when he flew in the Shuttle *Discovery*. Although he retired from NASA in 1964, he never gave up his interest in space, or a desire to return. In 1995 when he was a member of the Senate Special Committee on Aging, he realised that all the effects of weightlessness on an astronaut were similar to the effects on people on Earth as they grow older. He decided there would be a lot to learn by sending an older person up into space, and felt he would be the best person to go.

He managed to convince all the right people, from the government to NASA Administrator Dan Goldin that he could survive the medical. Goldin took some convincing, afraid of the reaction if something happened to Glenn during the flight. "My heart says yes, but my brain says no," Goldin admitted. Glenn did pass all the hurdles and climbed aboard Shuttle Flight STS-95 on 29 October 1998, returning back to Earth on 7 November. A number of scientific studies were conducted, a primary one being a sleep experiment. Certain physiological changes that occur in space also occur with aging; cardiovascular deconditioning, balance disorders, weakening bones and muscles, a depressed immune response, as well as disturbed sleep. Two subjects, one of them Glenn, were specially instrumented for sleep with electrodes for respiratory movement, ECG, sound, light, pulse oximetry, and nasal air flow.

This time his grandsons watched him take off. They were the same age as his sons were when they watched him become the first American to orbit the Earth.

During the flight Glenn had no problems with nausea, and when he landed he only experienced half the 6 g gravitational force he suffered in *Friendship 7*. Glenn said after the flight at first he did not feel too hot with slight problems keeping his balance when he turned his head quickly, but these effects did not last long.

1962 Mercury-Atlas 7: Scott Carpenter Makes Three Orbits

Scott Carpenter was a different astronaut. Most test pilots analyse the machine's performance and study how they arrived in space. Carpenter was more interested in what space was like and what could be out there. He wrote, "My mission, really, was to give the programme as hard a push as I could. I felt that John had been unduly handicapped during his flight, I was most anxious to let the capsule take care of itself and concentrate on things that he had not been able to do. I was to have more freedom to measure, study and observe events *outside* the capsule."[20]

Carpenter chose the name of *Aurora 7* because of the open manner in which the project was being conducted, "... for the

Mission Data: Mercury-Atlas 7	
Date:	24 May 1962
Craft name:	*AURORA 7*
Personnel:	Scott Carpenter
Duration:	4 hours 56 minutes 5 seconds
Features:	3 orbits of 88 minutes 32 seconds
Distance:	122,340.5 km
Apogee:	268.4 km
Perigee:	160.8 km
Speed:	28,242 kph
Spacecraft weight:	1,925 kg

Figure 2.31. The Operations Room at Muchea.

benefit of all as a light in the sky. 'Aurora' also means dawn – in this case the dawn of a new age."

Deke Slayton, originally scheduled for this flight, had just been told he was no longer on flight status because of an erratic rhythm in his heart called idiopathic paroxysmal atrial fibrillation. Although many medical panels agreed the condition was not serious and was not expected to affect his performance, NASA Administrator Webb didn't want to take any chances. He was grounded. Then he was sent as far as possible from America – he was Capcom at Muchea for Carpenter's flight.

After the smoothest countdown so far the launch was, "as near perfect as could be hoped for," according to Flight Director Chris Kraft. Once in orbit Carpenter found that weightlessness was, "such an exhilarating feeling that my report was a spontaneous and joyful exclamation: 'I am weightless!' Now the supreme experience

Figure 2.32. Deke Slayton was the Muchea Capcom for MA-7.

of my life had really begun. I could look for a thousand miles in each direction. I found it difficult to tear my eyes away and go onto something else."

Carpenter used a lot of fuel on his first night pass. "When I got over Australia, I was supposed to watch for some flares that were being set off on the ground to see how clearly they could be observed from space. Unfortunately, Australia was mostly covered by clouds and I did not see the flares."

He also suffered from excessive moisture and heat, the cabin temperature going above 37.7°C. Over America in the early morning light he tried to observe a balloon he ejected, but as it did not inflate he gave up. It stubbornly clung to the spacecraft though he tried to get rid of it, and trailed along with him until he re-entered the atmosphere.

Carpenter had different food to Glenn. He had three kinds of snacks: chocolate, figs, and dates with high protein cereals, plus some Nestlé's Bon-Bons. This food was processed into small blocks and coated with a glaze. Over Canton Island he tried a bite but his gloved hands and the helmet and microphone made it hard to put the food in his mouth. The food crumbled and particles drifted around the cabin, to become a nuisance and breathing hazard.

In the early stages of the mission, Flight Director Chris Kraft considered this was the most successful mission so far, but this was to change dramatically. Trying to race through experiments Carpenter bumped into switches and levers and had to reset them, as well as accidentally initiating the manual flying system. He used up a lot of fuel correcting mistakes and had to be reminded a few times to watch his fuel consumption, but he enjoyed his flight so much he recklessly used more fuel sightseeing. Then he had to drift for 77 minutes of the last orbit to conserve his fuel. When he should have been lining himself up to re-enter, he was busy turning the spacecraft to follow Glenn's "fireflies" which depleted his fuel even further. Then he found that the automatic flying system was misbehaving, and had to use a lot of manual control.

Carpenter advised he was behind in his pre-retro fire check list due to verifying Glenn's fireflies theory, then he suddenly found the automatic stabilising control would not hold the 34° pitch and 0° yaw attitude, so he fell back further. Going into re-entry Carpenter was facing a series of potential disasters – nobody knew how much fuel he had left, though they did know it was very low, the spacecraft was 25° off to the right of the correct angle for re-entry, he was three seconds late hitting the retrofire rockets which added another 24 kilometres to his landing error. Also the retrorockets did not fire to full power causing the main error in the landing position. On top of all that, the automatic control system was not steering the capsule properly.

Carpenter recalls, "At this time I noticed my appalling fuel state and realised that I had controlled retrofire on both the manual and fly-by-wire systems. I tried both the manual and the rate-command control modes and got no response. The fuel gauge was reading about 6%, but the fuel tank was empty. This left me with 15% on the automatic system to last about ten minutes to 0.05 g and to control the re-entry."[25]

While Carpenter was riding *Aurora 7* down the re-entry corridor, down in Mission Control Chris Kraft and the Flight Control team were seething with anger at the earlier irresponsible waste of fuel. Although they knew where Carpenter was, they wondered how he was going to survive re-entry if he had no fuel, also they weren't sure he had entered the heat barrier at the correct angle.

The airwaves were ominously silent when Carpenter should have been drifting down under the parachute with his signals booming in. On the television screens, Walter Cronkite was actually shedding tears as he announced that the capsule with the astronaut seemed to be missing. The next day newspapers wrote, "While millions of

Americans, glued to TV screens, wept, prayed, and chain-smoked, a rescue armada combed the seas," and, "In Central Station, New York, more than 8,000 people forgot their work, and stood, hushed, before a huge TV screen."

Quite oblivious to the consternation below him, Carpenter was dropping like a stone at 965 kilometres per hour into the atmosphere, but swaying wildly from side to side, building up to an arc of 270°. Fearing the capsule might topple over and start coming down nose first and the parachute might snag on its way out, he pushed the button to release the small drogue chute 1,524 metres above the planned height.

"I was still not frightened. It was a tight situation, and I was very alert. There just isn't time in a situation like this to get wide eyed and ask yourself, 'What'll I do now?' You don't worry because you can't – that *could* be sudden death. You just have to keep interested in what is going on and work your way out of it," he wrote.[20]

The 3,048 metres mark came and went with no big parachute. Carpenter says, "At about 9,500 feet (2,896 metres), I manually activated the main parachute deployment switch without waiting for automatic deployment." To his relief it sprang out and they soon settled into a pleasant drift through some clouds. He was heading for a landing 402 kilometres beyond all the ships and planes waiting for him, well out of their reach. He had heard Grissom on the radio say he was landing long, and it would be about an hour before anyone could get to him, but nobody was answering his calls because he was out of their radio range.

Aurora 7 landed in the Atlantic 201 kilometres north east of Puerto Rico. Carpenter remembers, "The landing was much less severe than I had expected. It was more noticeable by the noise than by the g-load. I was somewhat dismayed to see water splashed on the face of the tape recorder box immediately after impact. My fears that there might be a leak in the spacecraft appeared to be confirmed by the fact that the spacecraft did not immediately right itself.

"I knew that I was way beyond my intended landing point, because I had heard earlier the Capcom transmitting into the blind that there would be about an hour for recovery. I decided to get out at that time."[25]

Unable to find a leak, he struggled out of the hatch when the craft righted, and gazed at the ocean swells passing by. Some were as high as 1.8 metres but he felt quite safe. "I forgot to seal the suit, and deploy the neck dam. I think one of the reasons was it was so hot. After landing I read 105°F (40.5°C) on the cabin temperature gauge.

"I left the spacecraft, pulled the raft out after me, and inflated it, still holding onto the spacecraft. I climbed aboard and assessed the situation. Then I realised the raft was upside down! I climbed back onto the spacecraft, turned the raft over, and got back in.[25]

"And then I said a prayer: 'Thank you, Lord,' and relaxed for the wait. I have never felt better or happier in my life. I felt like a million dollars."[20]

Sucking a food bar and drinking water, Carpenter sat for a long time just thinking about what he had been through, the wonders of spaceflight, thinking how much he would like to return to the pleasure of weightlessness and the stupendous views from space but realising he could not share the experience with anyone. However he looked forward to telling others about it. "The first thing I saw in the water was some seaweed. Then a black fish appeared, and he was quite friendly. Later, I heard some planes. The first one I saw was a P2V, so I took out the signalling mirror from my survival kit. Since it was hazy, I had some difficulty in aiming the mirror, which is done by centering the small bright spot produced by the Sun in the center of the mirror. However I knew the planes had spotted me because they kept on circling the area."[25]

Two P2V reconnaissance planes and a Piper Apache found *Aurora 7* in 36 minutes, and three hours later a helicopter lifted a happy, though drenched, astronaut to safety aboard the USS *Intrepid*. Carpenter says, "I poked a hole in the toe of my left sock and stuck my leg out the window to let the water drain out of the suit.[25]

Meanwhile at Cape Canaveral the Flight Control team had quite different thoughts going through their heads – they were still recovering from the fright of a re-entry that was close to being out of control, and the portentous consequences of losing an astronaut. They were very, very angry. Carpenter would never fly for them again.

Flight Director Gene Kranz says, "He was very distracted during the course of his mission. We in Mission Control were just learning what this business was about at that time and, basically, a distracted astronaut that's behind in his flight plan is a danger to himself and the mission."

Many years later a mellower Chris Kraft summarised his feelings. "Carpenter was overcome with the moment, I suppose. I don't think he was versed in what a test pilot ought to do in flight. He acted like a kid on a lark. I don't think he was there as an engineering test pilot, he was there more as an interested and selfish observer."

Although publicly he was given the traditional astronauts' welcome home, Carpenter had lost the confidence of the mission flight controllers, so never did realise his dream of returning to space.

1962 Vostoks 3 and 4: The World's First Dual Spacecraft Mission

Nikolayev was the back up for Titov in Russia's second manned flight, who described him as the "calmest man in an emergency I have ever known." He once walked away from the flames of a horrific jet fighter crash, calmly writing notes on what was wrong with the plane. In November 1963, he married Valentina Tereshkova, the first woman in space.

This was the world's first manned dual mission, Vostok 4 flying within 4.8 kilometres of Vostok 3 and the first time two astronauts in separate spacecraft conversed in space. As they passed over Sydney on 13 August, the signals were picked up by various receivers operated by the CSIRO, radio hams, and the Island Lagoon tracking station at Woomera. A member of the CSIRO staff translated some of the conversation: "How do you hear me? How are things with you?" The reply was unintelligible, and the first voice said, "Excellent, I am very happy for you. It is much jollier that we are together."

Mission Data: Vostoks 3 and 4

Vostok 3
Date:	11 August 1962
Craft name:	FALCON
Personnel:	Andrian Nikolayev
Duration:	94 hours 22 minutes
Features:	64 orbits
Apogee:	235 km
Perigee:	180 km
Weight:	4,731 kg

Vostok 4
Date:	12 August 1962
Craft name:	GOLDEN EAGLE
Personnel:	Pavel Popovich
Duration:	70 hours 57 minutes
Features:	48 orbits
Apogee:	228 km
Perigee:	180 km
Weight:	4,731 kg

Communicating by radio, they came close enough to see each other, but then drifted apart and were unable to meet again because of the limited power of their propulsion systems. Both spacecraft also sent television pictures down from space which were broadcast around Europe, and later America. Korolev allowed both pilots to leave their ejection seats and float around the confines of their cabins. The cosmonauts ate packaged meals of chicken, veal, sandwiches, and fruitcake instead of sucking from a tube, and neither suffered any of the sickness that had plagued Titov. Although the mission was heralded as a successful rendezvous, Western observers did not fully realise that the Soviets were still a long way from the skills of meeting and docking their spacecraft in space. The rendezvous was accomplished more by careful launch timing than manoeuvring in space. At the end of the mission the two spacecraft landed 306 kilometres apart, within 6 minutes of each other.

1962 The Mercury Astronauts Elect a Boss
Deke Slayton was having a rough time with his heart condition and flight status, both the Air Force and NASA had decided to ground him, not even allowing him to fly with another pilot.

It was about this time that NASA needed a manager to organise the astronauts and run their office. Shepard thought if they had to have a manager, why not use Slayton – why call in an outsider? He approached Walt Williams, Wally Schirra, and Gus Grissom and they tackled Director Gilruth who agreed and Slayton became the coordinator of the astronauts' activities, "My first task was selecting a new group of astronauts."

1962 More Astronauts
Although out in the real world America was deep in the Cuban missile crisis, the NASA space programme was barrelling along with an influx of nine new astronauts on 17 September 1962, and by 26 January 1963 all were allocated special assignments under the supervision of Gus Grissom, already involved with the Gemini Program. Neil Armstrong, Frank Borman, Charles Conrad, James Lovell, James McDivitt, Elliott See, Tom Stafford, Edward White and John Young joined the growing band of astronauts. Except for Elliot See, who died in a plane crash, all these names were to become legends in the annals of spaceflight, and helped to fulfil President Kennedy's statement at Rice University on 12 September 1962: "We intend … to become the world's leading spacefaring nation."

1962 Mercury-Atlas 8: Walter Schirra Flies His Mission by the Book
Wally Schirra set out to show the world how a mission should be flown. Planned as an engineering evaluation mission, for the spacecraft's name he chose the mathematical symbol for summation, *Sigma 7*, to honour the immense engineering effort the project required. After Carpenter's irresolute trip and nail biting finish, Schirra flew his mission under strict control – his control. Where he felt he could use less fuel, he manoeuvred the spacecraft using manual mode. He would call down, "I'm in chimp mode right now and she's flying beautifully." The chimp mode was Schirra's reference to automatically controlled flights used for the chimpanzee flights.

Mission Data: Mercury-Atlas 8

Date:	3 October 1962
Craft name:	*SIGMA 7*
Personnel:	Walter Schirra
Duration:	9 hours 13 minutes 11 seconds
Features:	6 orbits of 88 minutes 55 seconds
Distance:	231,712 km
Apogee:	282.9 km
Perigee:	160.9 km
Speed:	28,256 kph
Spacecraft weight:	1,962 kg

After the fifth orbit, some kangaroos wandered into the grounds at Muchea, and the Canadian Capcom Gene Duret with all the Americans rushed out to meet these odd creatures from the land Down Under they had heard so much about.

Schirra's mission was such that 136 minutes of the flight was used to check drifting with no control. "I'm having a ball up here drifting," he told Mission Control. He conserved fuel so successfully that when he arrived at the re-entry phase he still had 78% of his fuel left, and that with a mission twice as long as Carpenter's. Mission Control were so pleased with this professional performance that one of the Flight Controllers made a point of coming on the link to the spacecraft and told Schirra, "Now that's what I call a *real* engineering test flight."

During the fourth pass over California, at 6 hours 8 minutes into the mission, Schirra and Glenn's capsule/ground conversation was broadcast live on radio and television for two minutes. Schirra broke out of spacespeak for a moment. "I'm coming toward you inverted this time, which is an unusual way for any of us to approach California, I'll admit. I suppose an old song *Drifting and Dreaming* would be apropos, but at this point I don't have a chance to dream, I'm enjoying it too much."

Sigma 7 landed 442 kilometres north east of Midway Island, and an accurate 8.2 kilometres from the prime recovery ship, the USS *Kearsarge*. When the parachute blossomed out Schirra quipped, "... that sort of put the cap on the whole thing."

Navy trained to his bootstraps, Schirra refused a helicopter lift:

Helicopter Pilot: "Astro. This is the Swiss pilot. The carrier is about three-quarters of a mile – closing."

Schirra: "OK, pilot. I think I would prefer to stay in and have a … a small boat come alongside and using your collar routine to support me and have a ship pick-up. Over."

Helicopter Pilot: "Roger, understand. You want small ship's boat. Will give them that right away."

Schirra called the *Kearsarge*'s captain, "Permission to come aboard, sir?"

"Permission granted," replied the captain, and Schirra waited for the carrier to come alongside. He had done everything in the Flight Plan: it was a 100% performance, and a

Figure 2.33. "Permission to come aboard, Sir?" The past sweeps in to recover the future as Navy pilot Wally Schirra waits for the longboat.

fine example to all other astronauts. Typically, Schirra laconically described his trip as, "A textbook flight. The flight went just the way I wanted it to." The only problem he had encountered was getting uncomfortably warm during the early orbits due to a sticking needle valve, but said he had been hotter sitting under a tent on Cocoa Beach.

Schirra took the first Hasselblad camera into space in this mission, to begin a close association between this camera firm in Göteborg, Sweden, and NASA. This collaboration later brought back the sensational images from space which have since become the icons of the century. The camera used in this mission was bought off the shelf of a local store.

1963 Mercury-Atlas 9: Gordon Cooper is Next

By this time, there were some criticisms of the huge effort being put into manned space flight so Cooper chose a name he felt was appropriate for the time. "*Faith 7* was the name I selected for the spacecraft, which performed so well for me until the electrical problem late in flight. I chose this name as being symbolic of my firm belief in the entire Mercury team, in the spacecraft which had performed so well before, and in God," Cooper explained.[22]

Up to this mission Chris Kraft had been the only Flight Director, but to cover the 34 hours he had to divide Mission Control into two shifts for the first time, and he became leader of the Red Team, and his deputy, Englishman John Hodge, became leader of the Blue Team.

Mission Data: Mercury–Atlas 9	
Date:	15 May 1963
Craft name:	*FAITH 7*
Personnel:	Leroy Gordon Cooper
Duration:	34 hours 19 minutes 49 seconds
Features:	22 orbits of 88 minutes 45 seconds
Distance:	878,946.5 km
Apogee:	267 km
Perigee:	161.4 km
Speed:	28,238 kph
Spacecraft weight:	1,814 kg

The laconic Gordon Cooper was a bit wild for some of the more conservative NASA people, beating up the launch centre in an F-102 jet with a supersonic display of daredevil antics only two days before his flight. The Associate Director for Project Mercury, Walt Williams, looked out of his window and saw Cooper scream past at eye level and reached for the phone to dial Deke Slayton to say he wanted Cooper brought under control – make him sweat a little.

So Slayton told Cooper that Williams was looking for him to ground him. Aware that they would happily put Shepard in his place without hesitation, Cooper settled down, and went on to be the first American to fly and sleep in space for a whole day. A relaxed Cooper delivered the goods as well, if not better, than Schirra. So relaxed, in fact, that he fell asleep on top of the rocket while waiting to be launched.

"I had thought I would become a bit more tense as the count neared minus one or two minutes, but found that I have been more tense for the kick-off when playing football. I felt that I was very well trained and was ready to fly a good flight," he said.[22]

After an "unbelievably correct" launch into a cloudless sky, at 8:04 am EST, Cooper noticed that when weightless in orbit the relationship with the cockpit changed. "You move up forward in the seat, regardless of how tight your straps are cinched. I did feel very distinctly that I was sitting upright. Most of the time I felt as if I were lightly floating.

A couple of times I felt almost as if I were hanging upside down because of the feeling of floating in the shoulder straps. Every time I dropped something I had the tendency to grab below it, expecting it to fall."[22]

In low humidity and cloudless areas such as the Himalayan Mountains, Cooper noticed he could detect individual houses and streets, some with smoke coming from their chimneys. He saw dust blowing off roads in the West Texas/Arizona and Himalayan areas, then could pick out the road, then when the light was right, an object which he thought was a vehicle. He also saw a steam train in Northern India by seeing the smoke first.

He found that the Earth had a sharp horizon, even at night. When there was no Moon, the Earth was darker than the sky – there was a difference between the two blacks, the sky had a shining black compared to the dull black of the Earth.

"One indication of my adjustment to the surroundings was that I encountered no difficulty in being able to sleep. When you are completely powered down and drifting, it is a relaxed, calm, floating feeling. In fact you have difficulty not sleeping. I found I was catnapping and dozing off frequently. Sleep seems to be very sound. I woke up one time from about an hour's nap with no idea where I was and what I was doing."[22]

After launching a flashing beacon into the night and trying another balloon experiment that also failed as he could not release it, Cooper went to a planned sleep in the tenth orbit, waking up over Muchea during the fourteenth orbit.

Due to a malfunction in the nineteenth orbit, suspected to be a water short-circuit, Cooper became the first American to return to Earth entirely by manual mode. Cooper says, "I had a malfunction associated with one of the control relays which eliminated my autopilot as well as my attitude indicators. Therefore I had to initiate retrofire, use window view for attitude reference, and control the spacecraft with the manual proportional system." To prove that a pilot was better than the machine, he landed even closer than Schirra, less than 7 kilometres from the USS *Kearsarge*. President Kennedy was impressed enough to announce, "One of the things that warmed us most during the flight was the realisation that however extraordinary computers may be, man is still the most extraordinary computer of all, and that we are still ahead of them."

Cooper was so efficient with his electricity, oxygen, and fuel, that the flight controllers called him a miser, and jokingly told him to stop holding his breath. Cooper was so busy taking pictures for various experiments such as zodiacal light and horizon definition he lightheartedly complained, "Man – all I do is take pictures, pictures, pictures!"

With the end of the Mercury Program the pilot had become less of a passenger, and more in control. Flight Director Chris Kraft reported that, "Man is the deciding element. As long as man is able to alter the decision of the machine, we will have a spacecraft that can perform under any known conditions, and that can probe into the unknown for new knowledge."

With the end of the Cuban missile crisis, Khrushchev turned his mind to baiting the Americans with more space achievements. What could be better than sending a woman into space? In response to this challenge four hundred women were screened for a single flight.

1963 Vostoks 5 and 6: The First Woman in Space

The Soviets continued to dazzle the world with new records, Bykovsky setting a new endurance record in space, and Tereshkova sitting on headlines around the world as the first woman in space. Bykovsky's flight had been planned to go longer, but with the detection of a large solar flare Korolev decided not to take any chances and terminated the mission on June 19.

Valentina Tereshkova was born in the village of Maslenikovo on 6 March 1937. Her father was a farm tractor driver and mother a weaver in a textile plant. She suffered the loss of her father fighting the Nazis when only six years old. While she was working at a textile mill in Yaroslav, she joined its aero club at the age of 21, and took up parachute jumping, completing nearly 150 jumps.

She dreamed of being a cosmonaut and, after the flight of Vostok 2, in 1961 she wrote a letter to Yuri Gagarin asking to join the cosmonaut core and was accepted. She nicely filled the image required by the Soviet officialdom: a qualified parachutist, and as a factory textile worker she had a genuine proletarian background.

On 14 March 1962 Valentina met the other finalists Tatiana Kuznetsova, V. L. Ponomareva, Irina Soloyeva, and Z. Yorkina, and these five girls endured exactly the same training as the male cosmonauts under the leadership of Yuri Gagarin. Initially the men were aghast at the thought of women anywhere near a flying machine, but once training began they came to accept female cosmonauts.

Mission Data: Vostoks 5 and 6

Vostok 5

Date:	14–19 June 1963
Craft name:	*HAWK*
Personnel:	Valery Bykovsky
Duration:	119 hours 6 minutes
Features:	81 orbits
Apogee:	235 km
Perigee:	159 km
Period:	88.3 minutes
Spacecraft weight:	4,731 kg

Vostok 6

Date:	16–19 June 1963
Craft name:	*SEAGULL*
Personnel:	Valentina Tereshkova
Duration:	70 hours 50 minutes
Features:	48 orbits
Apogee:	227 km
Perigee:	180 km
Spacecraft weight:	4,736 kg

Just over a year later, at 9:30 am on 16 June 1963 Valentina was launched into space in Vostok 6, with the call sign of *Chaika* meaning "Seagull". She flew around the Earth 48 times in three days, following Valery Bykovsky in Vostok 5, launched two days before, at one stage coming within 4.8 kilometres of each other. She supported a series of physiological, vestibular, and psychological tests for data on the female organism in space. She suffered a lot of disorientation and space sickness. Once during the flight, she did not respond to any of the ground control's signals for several hours, and they began to fear for her safety, but it turned out she was very tired, and had merely overslept. She suffered a bruised nose when landing. By the end of the mission she had racked up a distance of 2,000,360 kilometres and more flight time than all the American Mercury flights combined to that time.

After Tereshkova's flight Khrushchev was able to boast to the world about the high quality and equality of both the men and women of Russia. With the point proved the squad of female cosmonauts was quietly disbanded, and it was nineteen years before another Russian woman entered space.

Tereshkova married her training partner, Andrian Nikolayev, in one of the social events of the year. They produced the first child from two space travellers, a healthy

Figure 2.34. The first woman in space: Valentina Tereshkova. In her 48 orbits she racked up more hours in space than all the American astronauts in the whole Mercury Program, which had just been completed.

daughter, Yelena. Tereshkova reached the rank of engineer colonel in the Soviet Air Force, and also entered politics, becoming a Deputy of the Supreme Soviet in 1966, and a member of the Central Committee in 1972.

1963 The Mercury Project Comes to a Successful Conclusion

On 6 June 1963 Brainerd Holmes, Robert Gilruth, Walter Williams and Kenneth Kleinknecht met the NASA Administration of James Webb, Hugh Dryden and Robert Seamans in Washington to make a final decision on terminating the Mercury Project. President Kennedy left the decision to NASA. Although the astronauts wanted another flight Webb announced to the Senate Space Committee on 12 June 1963, "We will not have another Mercury flight".[9]

Dr Robert Gilruth, Director of the Manned Spacecraft Center, commented, "The exposure of man to zero gravity in these early manned flights was perhaps among the greatest medical experiments of all time. All the Mercury astronauts found the weightless state no particular problem. This finding was so fundamental and straightforward that its importance was missed by many medical critics at the time."[17]

1963: November A Sudden End to the Kennedy Era

Then came that fateful day, 22 November 1963 in Dallas, Texas, when a sniper denied President Kennedy the satisfaction of seeing his epoch-making decision to send a man to the Moon come to fulfillment. He was shot on his way to deliver a speech which included a short burst to defend the vast sums of money being expended on the nation's space programs: "This effort is expensive – but it pays its own way, for freedom and for America … in short, our national space effort represents a great gain in, and a great resource of, our national strength."

By the end of 1963, NASA began to feel more comfortable with Project Mercury successfully concluded; the worst of the upcoming Gemini Program's planning problems behind; and the master plan for Apollo complete. They also had a new President who they knew was sympathetic to their aspirations.

Although there was hardware for more Mercury missions, the next mission would have been MA-10 planned for three days, NASA Administrator Webb felt that Gemini was already scheduled to fly long missions in much more suitable spacecraft, and an accident at this point would only have delayed the whole Moon landing programme.

NASA management now felt quite ready and confident enough to build on the foundations of Mercury and move on to the next steps down the track to the Moon – the Gemini Program. Gemini planned to control man and his machines in the space environment; to change orbits; to get two spacecraft to meet and dock; and, very daring in 1963, to step out into the vacuum of space and take a walk!

3 The Gemini Program

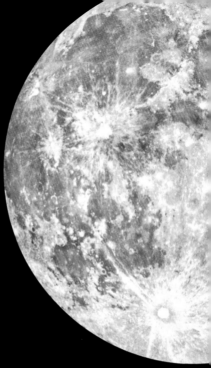

Now is the time to take longer strides – time for a great new American enterprise – time for
this Nation to take a clearly leading role in space achievement which in many ways may hold
the key to our future on Earth.

President John F. Kennedy State of the Union message to Congress, May 1961

Various names were used when Gemini was first proposed, from the first *Advanced
Mercury* to *Mercury Mark II*. Dr Robert Seamans, NASA Associate Administrator,
started a contest for a name in December 1961 and by 11 December, Alex Nagoya
from NASA Headquarters offered *Project Gemini*. He said, "This name, *The Twins*, seems
to carry out the thought of a two-man crew, a rendezvous mission, and its relation to
Mercury. Even the astronomical symbol 'II' fits the former 'Mark II' designation." Gemini
also symbolized the twin stars of Castor and Pollux.

Seamans had jokingly offered a bottle of Scotch whisky to whoever sent in the winning
name, and on 28 December the *Gemini Project* was chosen as the official title of the programme and it was announced publicly on 3 January 1962 and Nagoya won his bottle of
Scotch. On 1 November 1963 the word *Program* replaced the word *Project* to reflect its
responsibility of the programme as a whole, not just the spacecraft.[19]

George Mueller, the Associate Administrator for Manned Space Flight, defined the
goals of the Gemini Program:

"One of the important objectives of the Gemini Program was to determine the effects
of weightlessness during long duration missions. Another of the prime objectives was
the development of man's capability to step out into the nothingness of space and do
effective work. Finally, the ability to launch within a narrow time span, to rendezvous
with another craft in space, and to dock firmly with it is essential for manned operations
in space."[17]

To boost the spacecraft into orbit the Gemini Program chose the US Air Force's Titan
II two stage missile rocket burning "Arizona 50" fuel, a 50:50 blend of unsymmetrical-
dimethyl hydrazine and monomethyl hydrazine with an oxidiser of nitrogen tetroxide.

◀ **Figure 3.1.** "Get out in front where I can see you," called McDivitt before taking this view of Ed White,
holding his astronaut manoeuvring unit.

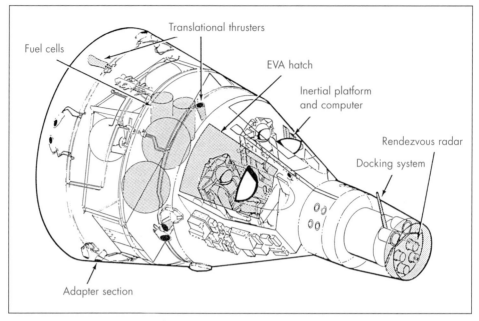

Figure 3.2. The Gemini spacecraft.

This fuel is hypergolic, which means that it ignites spontaneously when mixed. The non-explosive feature of these fuels allowed the Gemini spacecraft to use an astronaut ejection seat escape system instead of the rocket escape towers of the Mercury and Apollo systems. Twenty-seven metres tall, the Titan II first stage had a thrust of 195,048 kilograms, and the second stage had a thrust of 45,360 kilograms.

To the astronauts' disgust, the Mercury spacecraft had often been referred to as a "capsule" – a basic container to put an astronaut in – "a man in a can" – but with Gemini it matured enough to be called a spacecraft as it could now be manoeuvred around in space by a pilot. Using a system called the Orbital Attitude and Manoeuvring System (OAMS), 16 small thrusters could control the attitude of the spacecraft. For instance, when returning to Earth the Gemini pilot could steer the spacecraft 483 kilometres above or below the target, or veer up to 80 kilometres either side.

In the Mercury spacecraft most of the system components were packed around the pilot, scattered about to use all the limited space available. This system generated a maze of tubing, interconnecting cables, and mechanical linkages. To get at a malfunctioning component often other systems had to be removed, and all the systems laboriously tested again. James Chamberlin, chief designer of the Gemini spacecraft, introduced modular systems arranged in compact packages placed on the outside walls of the pressurised cabin for easy access. This allowed individual components to be removed without disturbing adjacent systems, and several technicians to work on different systems at the same time.

Carnarvon – a Space Town in the Australian Outback

A new tracking station was built for the Gemini missions at Carnarvon, on the mouth of the Gascoyne River. Carnarvon, once known as "the town too tough to die", was a real

Figure 3.3. The FPQ6 "C" Band radar's 9.1-metre dish antenna at Carnarvon. At the time, it was one of the most powerful and accurate radars in the world, with a maximum range of 51,000 kilometres. The data from this radar was the first confirmation that Houston had of the successful entry of a manned spacecraft into orbit.

Figure 3.4 (top). Trevor Housley monitoring the Digital Command System that sent instructions to the spacecraft, originally loaded from Mission Control.

Figure 3.5 (above). Dave Ricketts with the telemetry receivers that selected the engineering, technical and medical information received from the spacecraft.

frontier outpost, familiar with floods, cyclones, droughts, and bush fires. At the time it was an outback town of about 2,200 people, its broad main street originally designed to take Afghan drivers turning their big camel-drawn wool drays around. A line of scrubby trees down the middle provided the only relief outside from the scorching sun blazing down from the ever-blue sky. Clouds normally only appear in winter. Temperatures can go to 45°C, and it's one of the few places where even the oilcans rusted.

At the end of the main street a patch of blue water bordered by a beach and palm trees known as the Fascine, was the only concession to the city dwellers image of a tropical paradise. A picture of the Fascine was regularly used to lure unsuspecting wives and families of prospective highly trained station employees to take up residence in this remote town, particularly from cold, rainy, faraway England.

Station Director of the Carnarvon Tracking Station from 1968, Ray Jacomb said, "Carnarvon is a long way away from anywhere and it's only been those who wanted to come and who had enough individuality to go that far for a job that arrived to work at the station."

The first shipment of 15 tonnes of electronic equipment was delivered to Fremantle, near Perth, aboard the United States Lines freighter *Pioneer Reef*, and trucked the 965 kilometres north to Carnarvon, where the staff began to build the station, installing the equipment as it arrived.

1964 Carnarvon's Bush Telegraph System helps Gemini Mission

Carnarvon's first mission was a real Australian Outback story of the bush telegraph. It was Wednesday 8 April 1964 and the first unmanned Gemini trial, GT-I, was sitting on the launch pad ready to open the Gemini Program with a test of the structural integrity of the spacecraft and the launch vehicle. At Carnarvon the staff were still putting the finishing touches to the new station.

The author remembers that it was 10:22 pm local time – 1 minute 37 seconds to lift off. "We were standing by listening to the count, anxious to prove ourselves with our first mission. Everything was ready – we had all our mission information loaded, the equipment tuned up. Suddenly the line to Mission Control at Cape Canaveral went dead – at the time we didn't know what had happened, but we were cut off from the outside world by a lightning strike 105 kilometres south of the station."

Mrs Lillian O'Donoghue, the postmistress and operator of the weather station at Hamlin Pool at the southern end of Shark Bay, was roused up that night by a telephone call from the operator at Northampton, asking if she could contact Carnarvon. Using the bush telegraph – nothing more than a party line of telephones connected to the top strand of the local property fences, or in some places a line strung between the fence posts – Mrs O'Donoghue, who had only been in the job for four months, was able to speak to the operator 241 kilometres away in the town of Carnarvon.

The mission tracking data from Cape Canaveral was intercepted at Adelaide, and phoned through to the Postmaster General's Department test room in Perth. The Perth technicians then relayed the information to the technician at Mullewa, who established a phone patch through Northampton to Mrs O'Donoghue, and she and her husband then passed blocks of figures in half hour segments on to the Carnarvon operator from 10:30 pm until 3:45 am. From the Carnarvon telephone exchange it was a simple matter to get the information to the tracking station and the FPQ6 radar, a key element in the early phases of NASA launches from Cape Canaveral. It was 3 am before the PMG linesmen battled through driving rain and several washouts to get the normal landline operational again. After this episode a special tropospheric radio link was built between the station and Perth, and there were no more major communication breakdowns.

As Carnarvon had all the mission data already loaded before the lightning struck, the most important information was the time the spacecraft was launched and any changes. As Carnarvon wasn't officially completed and not a critical station for this particular mission, the launch went ahead, leaving Pad 19 on 8 April 1964. Sent into orbit faster than expected, the spacecraft ended up 34 kilometres higher than planned. One of the only two powerful FPQ6 radar's tracking at the time, Carnarvon followed the spacecraft over Australia until the mission was terminated after 64 orbits on 12 April, and came down in the South Atlantic.

The Russians Keep the Pressure On

On the other side of the Iron Curtain, the Russians were getting ready for their next move. After the flight of Vostok 6, the Soviets had originally planned to move into Soyuz flight testing. The name Soyuz, the Russian word for "Union", was probably chosen for the docking capabilities of the spacecraft, as well as recognising the Union of Communist States. The Russians had originally planned to send manned Soyuz capsules around the moon by the end of 1967, as unmanned tests had begun early that year.

With no space spectaculars ready to compete with the upcoming Gemini missions, Premier Khrushchev insisted something had to be done to keep the Soviet space achievements on the front pages and in the record books – if the Americans were putting two men up then the Russians would go one better and send three men into space!

So two discarded Vostok capsules were recovered from the scrap heap. Korolev was faced with the almost impossible task of jamming three men into a one-man capsule, mainly by removing the ejection seat and mechanism. By enlarging the main parachutes, and adding solid fuel rockets to fire just above the ground, a soft landing on ground was feasible. Renamed Voskhod, Russian for "Sunrise", these spacecraft were sent up to create more records for Russia, and stir the Americans.

Voskhod 1

On 12 October 1964, an A-2 booster sent Voskhod 1 up from Tyuratam with three men, Vladimir Komarov, destined to become the first person killed in space, Boris Yegorov and Konstantin Feoktistov. Feoktistov was the first engineer in space. Shot and left for dead by the Nazis during the war, he worked on the development of Sputnik and became second in command under Korolev in the design of the Vostok spacecraft.

Mission Data: Voskhod 1	
Date:	12 October 1964
Craft name:	*RUBY*
Personnel:	Vladimir Komarov
	Boris Yegorov
	Konstantin Feoktistov
Duration:	24 hours 17 minutes
Features:	16 orbits
Apogee:	409 km
Perigee:	177 km
Weight:	5,228 kg

While Komarov tested the new spacecraft's systems, Yegorov and Feoktistov conducted medical and geo-physical experiments, though the three cosmonauts were reported to be sick most of the time. In contrast to the Americans, the three cosmonauts whirled around the Earth in shirts and trousers, probably because there was no room to accommodate the bulky spacesuits. Also by this time the Russians had gained enough confidence in their life support systems. Komarov brought Voskhod 1 back to Earth with the first manned soft landing of a spacecraft on solid ground.

While they were in space, although only for a day, their leader, Premier Nikita Khrushchev, was suddenly tossed out of office and an era ended. Khrushchev had used the successful Russian space achievements to great political effect, blatantly pounding his fists at every opportunity to ram home to the world the technological superiority of the Soviet Union, and taunting the West to catch up. With the two big "K's" gone, the space race was never quite the same again.

Once in power, the new regime in the Kremlin, led by Premier Leonid Brezhnev, reorganised the whole Soviet space programme. They cancelled the Chelomei contract and in December 1964 Korolev was put in charge of the Russian space program, and given the green light to go ahead with the L1 project to put a cosmonaut on the moon using the N-1 booster. The success of this plan would have crushed any American goal to be the first on the moon.

NASA Prepares for the Gemini Manned Flights

The Goddard Space Flight Center in Maryland had organised a team of operational and engineering people to fly to the tracking stations around the world and train them all to the same standard, and also to evaluate their performance. So the first major event for the worldwide tracking network and Carnarvon was a visit from this simulation test team and a Super Constellation aircraft. While the aircraft flew back and forth over the station behaving like a spacecraft, the ground team put the station operational staff through searching exercises to train them in the procedures to follow when the real mission was in progress.

After delays because of unfavourable weather, including Hurricanes *Cleo* and *Dora*, the second Gemini launch test, called GLV-2, left pad 19 on 19 January 1965. To check its integrity, the spacecraft with a dummy crew was hurled 159 kilometres above the South Atlantic, to scorch back in to reach a temperature hotter than any mission so far. Picked up by the USS *Lake Champlain* after its 19 minute flight, the engineers found the spacecraft and its contents in good shape. Modifications to the Titan rocket's fuel distribution system dampened the violent oscillations that had been experienced just after launch, so now Gemini was ready to fly man. The first mission was planned to be Gemini III, with Gus Grissom and John Young.

Before we could raise Gemini III off the ground, the Russians startled the world – and us – again.

1965 Voskhod 2: The World's First Space Walk

Voskhod 2 supported the first walk in space. Pavel Belyayev and Aleksei Leonov were launched from Tyuratam at 10:00 am Local Time on 18 March 1965. The pilots began preparations for their spacewalk as soon as they were in orbit. Voskhod 2 was built with an airlock chamber so that Belyayev and the cabin equipment could remain in a normal atmospheric pressure while Leonov depressurised the airlock to enter space.

Mission Data: Voskhod 2	
Date:	18 March 1965
Craft name:	DIAMOND
Personnel:	Pavel Belyayev
	Aleksei Leonov
Duration:	26 hours 2 minutes
Features:	17 orbits
Apogee:	494 km
Perigee:	174 km
Weight:	5,139 kg

Some reports inferred an airlock was used because there was not enough oxygen and nitrogen stored on board to restore the cabin back to normal pressure. Belyayev activated the airlock while Leonov unstrapped himself from his couch and donned his space suit. Finally, after closing his visor, Leonov crawled into the airlock, and Belyayev closed and locked the hatch. The pressure in the airlock was reduced to check the integrity of the suit before then releasing all the air. He inched open the outer hatch, and floated out into he void.

With a light push he moved away from the spacecraft and first glanced down at the Earth, which seemed to move slowly past. Despite the thick glass of his helmet, he could see clouds to the right, the Black Sea below his feet, the Bay of Novorossysk, and beyond

Figure 3.6. Aleksei Leonov.

Figure 3.7. Leonov photographed during his spacewalk.

the coastline, the mountain chain of the Caucasus.

Pulling gently on his tether, he began to draw himself back to the spacecraft, then, pushing off again and turning around he moved slowly away again. He could see both the steady brilliance of the stars scattered over a background of black velvet, and at the same time the surface of the Earth. He could make out the Volga River, the snowy line of the Ural Mountains, and the great Siberian rivers Obi and Yenisei. He felt he was looking down on a great coloured map. The sun shone brilliantly in the black sky, and he could feel its warmth on his face through the visor.

He felt so good he had not the least desire to return back on board, even after he was told to get back in he floated away once more.

However, when Leonov did try to return to the airlock after a few minutes he was horrified to find he could not pass through the outer hatch as his suit had ballooned out from the internal pressure. What to do? Here he was floating along, looking down 161 kilometres to the Earth below, trapped out in space in his space-suit – and nobody around able to help! Belyayev was helpless inside the space-craft, only able to listen to his mate grunting with the exertion of fighting for his life. As there was only one spacewalking suit there was nothing he could do.

After a few minutes struggling desperately to wriggle into the airlock, with his pulse soaring to 168, Leonov tried letting the pressure of his suit drop down, but that didn't work. Desperate now, he tried again and brought it down to 26.2 kPa. Too sudden a drop, or more than a few minutes of high exertion at this pressure would have brought on a painful and probably fatal attack of the bends, but if he couldn't return to the cabin he would soon be dead anyway. With his suit now more flexible, he hooked his feet on the airlock edge and with the urgent desperation of a doomed man, elbowed and fought his way back in to the safety of the airlock. Leonov was out of the cabin for 23 minutes 41 seconds; 12 minutes 9 seconds of it outside the airlock. Belyayev reported that Voskhod 2 rolled and reacted every time Leonov hit or pushed himself off the spacecraft.

On the seventeenth orbit a fault developed in the spacecraft attitude system, refusing to line the spacecraft up for re-entry. Belyayev requested permission to take over manual control and they went around the Earth for another try. On the ground Korolev counted off the seconds to retrofire, which occurred over Africa. Voskhod 2 landed 3,219 kilometres away from Kazakhstan, the Ukranian target, way up among the thick forests of the frozen north near Perm in the Ural Mountains. Snow bound among the dense pine trees with little food and heating they spent the afternoon trying to keep warm in their spacesuits. As darkness fell upon them they lit a small fire for warmth, but Leonov spotted wolves eyeing them from the darkness, so they jumped back into the capsule, and spent the rest of the night huddled together listening to the growling and snarling of the wolf pack. Frozen stiff, they were very relieved when they peered out of the hatch the next morning to see a ski patrol sent to find them staring at the charred spacecraft, and their ordeal was over.

Manned Gemini Flights Get Under Way

The flight control teams spread around the world. Led by Danny Hunter, the team to conduct the mission from the Carnarvon station arrived and lodged at the local hotels. The Capcom was Pete Conrad, one of the second intake of astronauts, later to walk on the moon in Apollo 12. He recalled, "I remember flying in to Carnarvon the first time on MacRobertson Miller Airlines and the chap coming back and opening up the door and throwing my bags out on the red dirt and saying, 'See you on the way back, mate,' and off they went. That was at the end of about 54 hours of travelling.

"I checked into the hotel and asked the girl behind the desk did she have a reservation for a Lieutenant-Commander Conrad? She said, 'Upstairs on the second floor; take the first room that's made up and let me know the number on the way out!'"

1965 Gemini III: The First Manned Gemini Flight

Four years after the decision to start the Gemini Program the first manned flight was launched from Pad 19 at Cape Canaveral on 23 March 1965. Originally, Deke Slayton had chosen Alan Shepard to command this flight, but in May 1963 he went down with Ménière's Syndrome, excessive fluid pressure in his left inner ear which caused dizziness, ringing, vomiting, and sometimes dropped him to the floor. So he was grounded and joined Slayton organising the astronaut

Mission Data: Gemini III	
Date:	23 March 1965
Craft name:	MOLLY BROWN
Personnel:	Gus Grissom
	John Young
Duration:	4 hours 52 minutes
	31 seconds
Features:	3 orbits
Apogee:	224 km
Perigee:	158.5 km
Weight:	3,220 kg

corps, but he was determined to get back on flight status as soon as possible. He became somewhat moody and unpredictable and became known as the "Icy Commander" or "Smilin' Al" according to his mood. Gaye Alford, his secretary, would warn callers which mood to expect by hanging either a scowling picture or a cheerful portrait outside his door.

Slayton announced that Gus Grissom would command Gemini III, with John Young as his pilot just after the GT-1 trial was declared a success.

On launch day the daily afternoon sea breeze was blowing steadily over Carnarvon as the station staff were picked up by the little gray Commer buses dashing among the houses, and driven up to sun-baked Brown's Range. In the main T & C (Telemetry and Control) building, the staff escaped from the heat and sand outside, first grabbing a cup of tea or coffee and cooling off in the refreshing air-conditioning before spreading among the equipment to begin running through the final checklists.

Operations Supervisor Dick Simons settled down at the Operations Console, and began to bring the station together over the intercom. The author recalls, "We all settled down to our mission stations and listened to the countdown to launch on our headsets. I was surprised to receive an order from the Capcom to tune in to the short wave Voice of America, to give him the launch description in real time. The Voice of America fed us a continuous stream of information in great detail, much more than the occasional brief comment down our private SCAMA phone line from the Cape. At last we heard:"

The Cape: "You're on your way, *Molly Brown*."
Grissom: "Yeah, man."

Following his near-drowning episode in the second Mercury manned flight, Grissom was granted permission to call his spacecraft *Molly Brown* after a *Titanic* lady survivor and the stage show *The Unsinkable Molly Brown*. Grissom's spacecraft was the first and last one to be named in the Gemini Program. NASA officials were aghast at the choice of name, saying it lacked dignity, but when they heard his second choice was *Titanic* they promptly agreed to *Molly Brown*.[21]

At Carnarvon this mission began a brawl between the astronaut and the Flight Team leader for the position of Capcom, ending up with Pete Conrad taking the position from Danny Hunter, but it was the last time an astronaut communicated with the spacecraft during the Gemini Program.

In the initial orbit over Texas, Grissom fired two 38.5 kilogram rockets for 75 seconds to slow *Molly Brown* down by 15 metres per second and dropped it down into a nearly circular orbit. In the second orbit Grissom fired the rockets again, and shifted the plane of their orbit. Both manoeuvres were firsts for a manned spacecraft. "This was a big event, really a big event," Grissom said later.

Another event that seemed minor but became big with repercussions reverberating all the way up to Congress, was John Young's corned beef sandwich from Wolfie's delicatessen at Cocoa Beach.

Young said, "It was no big deal – I had this sandwich in my suit pocket. The horizon sensors weren't workin' right so I gave this sandwich to Gus so he could relax – there was nothing he could do in the dark to make that thing work, until we got back into the daylight."

"It negated the flight's protocol," thundered the doctors. "The crumbs could have got into the machinery," complained the engineers. "NASA has lost control of the astronaut group," boomed hostile voices around the floor of Congress. Grissom later admitted that the sandwich was one of the highlights of the mission for him.[21]

In the third orbit Grissom completed a fail-safe plan with a 2½-minute burn that dropped the spacecraft perigee to 72 kilometres to make sure of re-entry even if the retrorockets failed to work. This was added to the flight plan to protect the Gemini 3 crew against being stranded in space in case of a failure of the retrorockets, prompted by Martin Caidin's movie *Marooned*.

Just before landing, Grissom threw a landing attitude switch, and *Molly Brown* snapped into the right angle to land, pitching both men into the window and breaking Grissom's faceplate, before they dropped into the Atlantic, 111 kilometres from the USS *Intrepid*. The Gemini spacecraft produced less lift than predicted so landed about 84 kilometres short of the target. As they landed the spacecraft was dragged along nose under water by the parachute. All Grissom could see through the window was sea water, and with his Mercury flight still fresh in his mind, he released the parachute, but this time was not going to "crack the hatch", so the two astronauts suffered a miserable 30 minutes sealed in a "can" that was getting hotter by the minute, and being tossed around by the seas.

Young: "It was a really good test mission. Gus performed more than 12 different experiments in the three orbits – he did a really great job – I don't think he really got enough credit for the great job he did. He proved that the vehicle would do all the things needed to stay up there for fourteen days. We changed the orbit manually, the plane of the orbit, and we used the first computer in space." John Young's hometown of Orlando, Florida, showed their support for him by presenting him with a 18 metre long telegram signed by 2,400 residents.

The Manned Spacecraft Center at Houston, Texas, Comes on Line

Gemini III was the last time that the Mission Control Center at Cape Canaveral was used. The familiar flight controllers' voices on the network saying, "This is the Cape," was replaced by, "This is Houston," as the new Manned Spacecraft Center at Houston, Texas, came on line for the Gemini IV mission. Chris Kraft had to add two more Flight Directors and flight teams to cover the long Gemini missions, and Gene Kranz with his White Team and Glynn Lunney with his Black Team joined the Red and Blue teams.

1965 Gemini IV: The Americans Walk in Space, too

Originally, the first Gemini EVA (Extra Vehicular Activities), commonly called a spacewalk, was scheduled for Gemini VI, but because of the Russian Leonov's walk, and the American spacewalk suits and systems having been announced as ready for use ahead of schedule, NASA began looking at bringing the spacewalk forward. Ed White was ideal for the first American attempt to walk in space; a fitness fanatic and superb athlete, he just missed out on the US team for the 400

Mission Data: Gemini IV	
Date:	3–7 June 1965
Personnel:	James McDivitt
	Edward White
Duration:	97 hours 56 minutes
	12 seconds
Features:	66 orbits, 62 revs
	EVA of 21 minutes
	by Ed White
Apogee:	296.1 km
Perigee:	159.4 km
Weight:	3,538 kg

metre Olympic hurdles by 0.4 of a second. Some people at NASA were still not going to be pushed: "… we shouldn't be putting guys in a vacuum with nothing between them but the little old lady from Worcester and her glue pot," they warned, referring to the

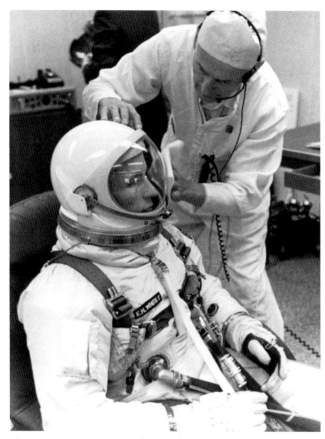

Figure 3.8. On Gemini IV launch day, Suit Technician Joe Schmitt gives a final polish to Ed White's visor.

seamstress at the suit manufacturing plant of David Clark Company, in Worcester, Massachusetts.

Although Grissom and Young had tried unsuccessfully to depressurise the cabin and open the hatch during one of their Gemini III simulations, the NASA hierarchy did reluctantly agree to have White stand up on his seat, but after Leonov's walk President Johnson is supposed to have snarled, "If the guy can stick his head out, he can also take a walk. I want to see an American EVA."

At 10:16 am USEST on 3 June 1965 the 27 metre tall Titan II rocket thrust the Gemini IV capsule into the sky in front of the first international television audience, the Europeans watching it through the *Early Bird* satellite.

Once in orbit, McDivitt and White tried to catch up with their discarded booster rocket. As the programme was still learning orbital rendezvous techniques, McDivitt, who was only eyeballing his manoeuvres, gave up when their fuel quickly ran down to half, and they immediately began preparations for the first American spacewalk.

On the third pass around Carnarvon, the astronauts were finding conditions very difficult, and McDivitt called to the Carnarvon Capcom, Ed Fendell:

McDivitt: "Listen, you might advise Flight that we are running late on this thing, there's a lot to do and we are having trouble keeping track of all these things. I'll give you a blood pressure as soon as I get around to it."

Fendell: "Full scale on your blood pressure."

McDivitt: "I don't think you got a good blood pressure – the bulb popped off there."

Fendell: "Gemini IV you are *go* for EVA and decompression. Disregard the blood pressure unless you have got some minutes."

McDivitt snapped back: "We don't have any time at all. We're really pressed here."

Fendell: "We're not going to say anything here on the ground. If you need anything we're here waiting."

McDivitt: "OK. Listen, just advise the Flight Director we're runnin' a little late – we might not be ready at Hawaii."

Fendell: "OK. He's ready – he knows that. Houston advises you can use any attitude you like for your extra vehicular activity."

On the next pass over Carnarvon McDivitt checked:

McDivitt: "I understand we have a *go* to start decompression, is that right?"

Fendell: "That's affirm. A *go* for decompression and a *go* for EVA."

McDivitt: "Roger. We expect to be out by the time we get to … near Hawaii."

Gemini IV was 193 kilometres above Hawaii when the smooth Public Affairs voice announced, "This is Gemini Control. Four hours and twenty four minutes into the mission. The Hawaii station has just established contact and the pilot, Jim McDivitt advises the cabin has been depressurised. It is reading zero. We are standing by for a GO from Hawaii to open the hatch. … White has opened the hatch. … He has stood up. McDivitt reports that White is standing on the seat."

The author remembers, "At Carnarvon we were all getting ready for the next pass, but hanging onto every word coming down the voice channel from Houston. To us this first spacewalk was one of the supreme moments of the Gemini Program, and we were agog to hear how it was going, and what we would find when they came up over our horizon."

While the spacecraft was travelling between Hawaii and Mexico, White eased himself out of the hatch with the manoeuvring unit, and found himself hovering above the

Figure 3.9. The astronauts opened the hatch of Gemini IV and Ed White soared into space to become the first American spacewalker.

spacecraft. A glove he had left on the seat seemed to acquire a mind of its own – it rose off the seat and gently drifted out of the hatch after him to waft off into space.

White remembers, "There was absolutely no sensation of falling. There was very little sensation of speed, other than the same type of sensation that we had in the capsule, and I would say it would be very similar to flying over the Earth from about 20,000 feet. You can't actually see the Earth moving underneath you. I think as I stepped out, I thought probably the biggest thing was a feeling of accomplishment of one of the goals of the Gemini IV mission. I think that was probably in my mind. I think that is as close as I can give it to you."

White propelled himself down to the nose of the spacecraft, then back to the adapter end, but soon ran out of fuel, and reported, "The manoeuvring unit is good. The only problem is I haven't got enough fuel. I've exhausted the fuel now and I was able to manoeuvre myself down to the bottom of the spacecraft and I was right on top of the adapter. I'm looking right down, and it looks like we are coming up on the coast of California, and I'm going in slow rotation to the right. There is absolutely no disorientation association."

McDivitt observed, "One thing about it, when Ed gets out there and starts whipping around it sure makes the spacecraft tough to control."

White then began to use the umbilical tether to move around, and explained, "The tether was quite useful. I was able to go right back where I started every time, but I wasn't able to manoeuvre to specific points with it. I also used it to pull myself down to the spacecraft, and at one time I called down and said, 'I am walking across the top of the spacecraft,' and that is exactly what I was doing. I took the tether to give myself a little friction on the top of the spacecraft and walked about three or four steps until the angle of the tether to the spacecraft got so much that my feet went out from under me. I also realised that our tether was mounted so that it put me exactly where I was told to stay out of."

While McDivitt sat at the controls keeping the spacecraft as steady as possible with its nose pointing down at the bulk of the continental United States spread out below, White moved around while they both took photographs and discussed the view.

Capcom: "Take some pictures."
McDivitt: "Get out in front where I can see you again."

In Houston, Flight Director Christopher Kraft was beginning to look anxiously at the time. The Flight Plan called for a space walk of 12 minutes, and it was already well past that with no signs of White returning.

McDivitt: "They want you to get back in now."
White: "I'm not coming in – this is fun!"

Gus Grissom, the Capcom at Mission Control was trying to get White to return, and curtly ordered, "Gemini IV – get back in!"

"But I'm just fine," returned White, in high spirits.

Commander McDivitt snapped back, "No, get back in. Come on. We've got three and half more days to go, buddy."

"I'm coming." White's boots thumped on the spacecraft as he reluctantly worked himself to the top of the capsule, handed back the camera, and again stood on the seat. Savouring the moment he stood briefly on the seat, looked at the stunning view, and sighed, "It's the saddest moment of my life."

Grissom queried what was happening, and McDivitt replied, "He's standing in the seat now, and his legs are down below the instrument panel. He's coming in."

White had been outside the spacecraft for 21 minutes.[31]

After White struggled to get back into his seat in the spacecraft they closed the hatch and grabbed the ratchet handle to secure it. To White's shock it failed to catch, so the fastenings had to be manoeuvred into place by hand and secured, McDivitt holding him down. They both collapsed back into their couches, physically exhausted, with sweat streaming into their eyes, and fogging their faceplates. White became so overheated from the struggle to get back in it took him a few hours to return to normal. The EVA equipment was supposed to be dumped in space to give the astronauts more room in the cramped cabin, but the hatches were never opened again so it all had to be carried around for the rest of the mission.

Figure 3.10 (left). "Gemini IV, *get back in!*" A vexed Gus Grissom orders a reluctant Ed White to finish his spacewalk...

Figure 3.11 (below). ... but White answered, "I'm not coming in – this is fun!"

Figure 3.15. Tom Stafford (left), with Mission Commander Wally Schirra settled down in Gemini VI-A before blasting off from...

Figure 3.16. ... the Kennedy Space Center's cluttered landscape of launch structures for a rendezvous in space with Gemini VII on 15 December 1965...

Houston: "Carnarvon, can you hear any music on HF?"
Kundel: "Yes."
Houston: "What are they playing now?"
Kundel: "*I'll Be Home For Christmas*."
Houston: "Congratulations – you have won a free trip back to the USA!"
Kundel: "All flight controllers punch up HF and help me!"

While Borman and Lovell were swinging around the Earth, eating and sleeping, and the ground networks fell into a monotonous routine following them, the launch crews at Cape Canaveral were feverishly preparing for the second launch. Forty-five minutes after Gemini VII had departed, the Gemini VI-A booster rocket was on its way to the launch pad. There were only nine days to go to launch.

To everybody's surprise the launch crews were so quick and efficient they actually finished in eight days, so Schirra and Stafford were sitting on top of the rocket again by 12 December, and the count proceeded smoothly to lift off. It was critical that they lifted off precisely on time with no hold in the count, for every 100 seconds delay Gemini VI-A would have to wait for Gemini VII to go around the Earth again. The two main engines ignited; the rust coloured smoke enveloped the launch pad; the clock started on the astronauts' instrument panel – *but Gemini VI-A didn't go!*

In less than 2 seconds the main engines promptly shut down. Couched in the launch position, Schirra and Stafford steeled themselves for take off, holding onto the D handles to eject in case of an emergency, when both astronauts suddenly realised they *were* in an emergency. The count had passed "0", the clock and some of their instruments were telling them they were on their way, yet there was no movement, their seats still felt securely anchored to the ground.

Stafford recalled, "We had all the lift off signals, and Wally and I knew in the seat of our pants we hadn't lifted off – also Al Bean never called we had lift off. We never had a simulation like this – we had shut downs but not with all the signals saying we had lifted off!"

There were three seconds for Schirra to make up his mind whether to pull the ring and eject to safety, or stay and take the chance that they might be enveloped in a deadly fireball from a disintegrating rocket. Two other men in the ground crew, not toughened test pilots, also had to make the same decision. All *three* men arrived at the same decision within those three tense seconds – sit tight! Schirra calmly announced, "Fuel pressure is lowering." He knew the rocket had not moved, and he knew the clock was wrong.

Kenneth Hecht, chief of the Gemini escape systems was surprised when they didn't eject, as they should have if the ground rules had been strictly followed. Had the rocket risen just a few centimetres, the engine shutdown would have dumped 123 tonnes of volatile fuel encased in a fragile metal shell crumpling back to Earth and the resulting holocaust could have engulfed the helpless astronauts.

The Titan II rocket and Gemini spacecraft were saved for another attempt thanks to the unemotional judgment of trained people, trained by endless simulations to prepare for just such an emergency. They knew exactly what to do – or was it a reluctance to face the enormous acceleration of the ejection seat?

Nobody had actually tried the escape system – what would the sudden 20 g acceleration do to a pilot? Nobody was quite sure. Schirra commented, "If that booster was about to blow … if we *really* had a lift off and settled back on the pad, there was no choice. It's death – or the ejection seat!"[32]

The author recalls, "Initially the cause of the rocket shutdown was suspected to be an electrical plug falling out prematurely, and this explanation was handed out to the media.

Figure 3.17 a. ... where they found Frank Borman and Jim Lovell cruising along above the Earth's brilliant blue atmosphere. Stafford took this picture of Gemini VII when they were 11 metres apart.

Figure 3.17 b. Fisheye view of the Gemini VII cockpit.

During a visit to Carnarvon after the mission Wally Schirra told us that they found out later another reason was a simple procedural error. A set of five small plastic caps had to be removed from the Titan booster's gas generator. The mechanic announced he had removed the caps, an observer checked he had removed the caps, and the message was relayed back that the caps were off. The procedures had not specified the number of caps to be announced – the mechanic had only removed four of five caps!"

Zooming past 297.7 kilometres above the launch pad, Borman and Lovell had watched the events unfold through a small telescope, and after being told the bad news, Borman called down, "We saw it light up; we saw it shut down."

At last, hoping the problems were sorted out, the astronauts climbed into the capsule for the third time on 15 December and at 8:37 am were launched to a successful rendezvous with their friends in Gemini VII, now rather travel weary after 11 days in space. Above Tananarive, Malagasy Republic, Lovell and Borman saw a brief glimpse of Gemini VI-A, and began to put on their suits ready for their visitors. It took six hours of manoeuvring for Schirra and Stafford in Gemini VI-A to reach Gemini VII. At a distance of 434 kilometres the spacecraft radar was able to lock on.

1965 Americans Greet Americans in Space

Schirra exclaimed, "My gosh, there is a real bright star out there. That must be Sirius." It wasn't, it was Gemini VII reflecting the sun's rays from 100 kilometres away.

The author remembers, "Down on the ground at Carnarvon we were finding our resources stretched to the limit tracking two talking spacecraft at once – the first time they passed over the station 4 minutes apart, the second time they had closed the gap to 2 minutes, and by the third pass Gemini VI-A was only 59 seconds behind. There were now conversations from two spacecraft to us, conversations between them, and Houston trying to pass messages to everybody. Our equipment had not been designed to cope with this."

Stafford said, "The rendezvous went a lot like the simulations – we went right on the money. Maybe 80 miles out, I had the critical manoeuvre and the terminal phase calculated before the IBM computer on board did or the ground did."

Lovell added, "Borman and I were looking for Gemini VI. We strained our eyes, looking where we thought they should be, when Borman suddenly nudged me and said with a broad grin, 'There they are.' They were less than two miles from us, looking like a glittering star, getting bigger all the time. They came straight towards us, until it looked as though we were going to hit each other."

Borman said, "At a distance of a few yards they suddenly seemed to stop, sitting there motionless – we were flying along nose to nose. Then we lay side by side, as Gemini VI slowly turned around us." Both Borman and Lovell were astonished to see great 12-metre tongues of flame belch from Gemini VI-A's thrusters as they braked.

Schirra and Stafford were very pleased with their efforts – the rendezvous had only used 51 kilograms of fuel so they still had 62% of their fuel left. They were also excited by their ability to manoeuvre their spacecraft with such precision – they could see no problems with docking two spacecraft together.[32]

Flight Dynamics Officer Jerry Bostick recollected, "I redistributed the flags we had got for Gemini VI and on 15 December 1965, two days before the birth of my daughter Kristi, when the two spacecraft pulled together, cigars were lit, cheers went up, and the flags were waved vigorously. At a time when some in the country were burning the US Flag, we waved it with great pride. We were proud of America and we were proud of what we were doing in the American Space Program."

Schirra edged his spacecraft to 6 metres, then 0.3 metres from Gemini VII. After flying nose to nose, Schirra moved Gemini VI-A slowly around Gemini VII while both crews

Figure 3.18a. "B—— VISITORS! Quick pretend we're not home!!..."

Figure 3.18b. Happiness is a successful rendezvous. Chris Kraft (left) shared his happiness with astronaut Gordon Cooper and a well satisfied Dr Robert Gilruth when Gemini VIA and VII came within a few metres of each other in the first American rendezvous in space.

took photographs. The astronauts found they could see each other quite well. Lovell in Gemini VII asked Schirra, "Can you see Frank's beard, Wally?"

Schirra answered, "I can see yours better right now."

They kept station for $5\frac{1}{2}$ hours during which Stafford explains, "We flew around and took all types of pictures, looked at things, talked to each other, and described all those cords hanging off the back end of Gemini VII." Both crews were surprised at all the dangling pieces hanging off the back of their spacecraft.

Schirra was worried about visibility on the night side of the Earth, but found that docking lights, pen torches, and cabin lights were clearly visible in the dark. After six hours of flying together, Gemini VI-A pulled about 14 kilometres away and both crews turned in for a sleep period before Gemini VI-A set out to return to Earth, Schirra remarking, "Really a good job, Frank and Jim. We'll see you on the beach," and the visitors splashed down in the Atlantic 21 kilometres from the carrier USS *Wasp*, leaving Borman and Lovell to wrestle with their cantankerous fuel cell.

After the departure of Gemini VI-A, life in the last stages of the marathon mission seemed very empty to Borman and Lovell, and thoughts would often turn to home. The Gemini VII astronauts would have been quite happy to finish the mission

Figure 3.19. Up close and personal with the Atlantic Ocean. Tom Stafford wonders if the sea isn't getting up too close as the spacecraft heaves over a swell. Gemini VI-A had just landed safely after a successful rendezvous with Gemini VII, still orbiting above.

there and then, as all the objectives had been met, but Chris Kraft, on his last mission as a Flight Director, cunningly coaxed them along to the end of the official flight plan.

Kraft recounted, "It wasn't easy. They seemed very melancholy after Gemini VI left them. I talked them into it because as we got away from the continental United States the prime landing sites then became very sparse and even when we had them it would take us several hours to get a destroyer there. They didn't want to go by those primary landing sites of the United States unless they were pretty sure we were going to be able to go for another fourteen revolutions. I met with them on a private conversation and told them that, and they were happy to hear that. I convinced them they should stay. What we were trying to do was match the longest period that we could think of for an Apollo flight, so we wanted the fourteen days and George Mueller was pretty insistent that we do that. As a matter of fact, I suspect that if George Mueller hadn't been as insistent as he was, we might have terminated that flight earlier."

All through the fourteen days the two sang Nat King Cole's *Put your sweet lips closer to the phone*, and to help pass the time they read Mark Twain's *Roughing It* and *Drums along the Mohawk* by Walter Edmonds, both chosen because they had nothing to do with space! Neither finished their books.

Gemini VII's parachutes finally dumped the tired spacecraft and its two travel stained occupants into the Atlantic at 9:05 am USEST on 18 December in time for Christmas, after a record distance of 8,296,390 kilometres, taking them 206 times around the Earth in 14 days. Surprisingly, in many ways the two astronauts were in better shape than their colleagues Conrad and Cooper after eight days in Gemini V, mainly put down to a better sleeping schedule. Conrad and Cooper slept alternately, and were regularly woken up by the duty astronaut conversing with the ground. In Gemini VII both astronauts slept at the same time, so were able to get a more restful sleep.

Back on earth Borman felt a little dizzy, but Lovell felt fine, though he had to admit the trip "… was like spending fourteen days in a men's room."

Agena was not responsible for any of the Gemini VIII problems, and the focus of attention turned to the spacecraft itself.

Hauled back to its birthplace at McDonnell's plant at St Louis, Missouri, for a month of intensive investigations, engineers decided that control thruster number 8 had stuck in the firing position due to an electrical short circuit, sending the Gemini spacecraft spinning out of control. McDonnell found that although the control switch was off, power was still available at the thrusters so they modified the circuit to stop any possibility of the thrusters firing with the switch off.[32]

Despite their narrow escape, the two astronauts were still enthusiastic about returning to space, and went on to command Apollo flights with distinction.

Figure 3.22. Gemini VIII – Surprise ending. Scott and Armstrong were disappointed to find themselves in the sea alongside the destroyer USS *Leonard F. Mason*, off the coast of China, after only $10\frac{1}{2}$ hours in space.

1966 **The Troublesome Gemini IX**

Gemini IX was plagued with troubles. First of all the prime crew of Elliot See and Charles Bassett were killed on 28 February 1966 on a visit to the McDonnell plant at St Louis to inspect their Gemini capsule and run through a practice in the factory simulators. Tom Stafford and Eugene Cernan also went along as the back up crew.

Cernan recalls, "We left Houston and we flew on their wing in two plane formation all the way up to St Louis. The weather at St Louis was marginal – the clouds were ragged, about four or five hundred feet above the ground in places. We made a two-plane approach to the runway, and Elliot See didn't quite get down to the minimum under the cloud early enough in his approach, and he was fast. We went over the field and circled back around the field to do another approach in two plane formation – we came out of the clouds and there was the field and Elliot chose to try and dive down quickly while he could see it and land, which surprised us. Tom Stafford and I decided to stand off and make another normal instrument approach. Once they broke away from us we never saw them again. The tower switched us to another frequency and we went out and flew another approach as a single airplane.

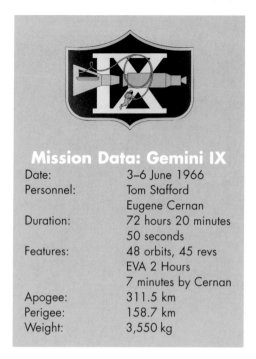

Mission Data: Gemini IX

Date:	3–6 June 1966
Personnel:	Tom Stafford
	Eugene Cernan
Duration:	72 hours 20 minutes
	50 seconds
Features:	48 orbits, 45 revs
	EVA 2 Hours
	7 minutes by Cernan
Apogee:	311.5 km
Perigee:	158.7 km
Weight:	3,550 kg

"Of course, we didn't know it at the time but See and Bassett saw the field and put their nose down and made some sharp turns. They couldn't make the first runway they saw, so they circled around in a very tight turn in a very high performance airplane and they tried to make another runway at very low altitude, and it looked to us like it stalled out and they crashed into that building at McDonnell."

See and Bassett's wing tip hit the roof of the very building they were due to visit and they were both killed as the plane cartwheeled over the edge into a parking lot and exploded, injuring 14 factory workers.

This accident began the astronauts' musical chairs that eventually ended up determining the first crew to land on the Moon.

Stafford and Cernan now became the prime crew, and all the crews and their back ups then changed, allowing Buzz Aldrin to fly in Gemini XII, and giving him spaceflight experience before the Apollo missions began, opening the door for him to join the Apollo 11 crew.

The Agena target was launched from pad 14 on 17 May 1966 but one of the Atlas rocket engines swivelled over to its limit, flipped the whole vehicle over, and sent it on a nose dive 257 kilometres out in the Atlantic. Next Stafford and Cernan heard in their earphones, "We have lost our bird – the mission is scrubbed."

A simplified docking vehicle called an ATDA, or Augmented Target Docking Adapter, which the astronauts nicknamed "The Blob", was hurriedly launched on 1 June. Seven minutes later it was safely in orbit, but there were signs that the fibreglass launch shroud might not have been jettisoned properly. All was not well on the ground either, as Stafford and Cernan had to suffer three attempts to

Figure 3.23. Charles Bassett and Elliot See.

get up and chase the ATDA due to equipment glitches. On the last attempt Cernan spotted a sign hung on their elevator saying that the down capability of the elevator had been disabled! Stafford, dubbed the Mayor of Pad 19, complained, "Frank Borman and Jim Lovell may have had more flight time, but nobody has had more pad time in Gemini than I did!" He had sat and waited to go after targets six times.

Flying in darkness over the eastern Pacific Ocean on their third orbit, the Gemini IX astronauts could see a flicking light way ahead – the strobe on the ATDA. When they eventually sidled up to it they were disappointed to find the nose shroud was still attached by a steel band, looking just like an alligator's jaws. Stafford reported through the Hawaii tracking station, "We've got a weird-looking machine here. Both clam shells of the nose cone are still on, but they are wide open ... it looks like an angry alligator out there rotating around."

Following conferences on the ground the idea originally proposed by Buzz Aldrin that Cernan should spacewalk out and snip the offending band was rejected partly because of the possibility of the band snapping back and puncturing Cernan's spacesuit. Gemini IX then backed off and tried to rendezvous without the help of radar, finding it more difficult than they expected using just their eyes, a computer and a pencil and pad of paper. Becoming tired they had a break and tried to rest before trying a third rendezvous

Figure 3.24. The angry alligator.

to simulate an Apollo Command Module trying to rescue a Lunar Module stuck in a lower orbit.

By now they were feeling very weary and felt they couldn't hack a spacewalk now, so Stafford advised the ground, "Right now we're both pretty bushed, we've been busier than left-handed paper hangers up here and it might be better for both of us to knock off for a while."

Mission Control puzzled over this new phenomenon of their flight crew refusing duty but reading between the lines that perhaps they might be close to total collapse, agreed, so Capcom Neil Armstrong called back, "It's the ground recommendation that we postpone the EVA until the third day. Would you agree with that?"

1966 **First to Walk Around the World in Space!**

After a ten-hour rest period and four hours preparation Cernan opened the hatch and became the third man to step out into space. He found after looking through the small spacecraft window the expansive view was: "Wow!" Seeing the tops of thunderclouds, the blue oceans and seas with tiny "vee" shapes of ships' wakes, vast mountain ranges, and the Mississippi River snaking down to New Orleans pass under his boots was a magnificent feast for the senses. Cernan felt no "space euphoria" he had been warned about. "Gemini IX was a very ambitious EVA. We had only had one Russian EVA of 12 minutes, and we didn't know much about his episode, and Ed White had been out there for 21 minutes. All he had to do was float out there and see if this gun would drag him around. He did a tremendous job, but the problem was we all totally overlooked the importance of Newton's Laws of Motion – for every action there is an equal and opposite reaction. None of us among the astronauts, or on the ground, really took that into account with any great seriousness so the major goal and focus for my EVA was to fire the AMU, or Astronaut Manoeuvring Unit. By the way, at that time it had hot gas rocket engines on it so I had to wear woven steel pants over my regular space suit to keep from burning holes in it. The only ones ever worn in space! Crazy! We would never do that today. Today we use cold nitrogen gases.

"The state of the art of spacesuits at that time were somewhat primitive – there were no arm bearings, the mobility was very difficult, it was pressurised to $3\frac{1}{2}$ psi. It was like a suit of plaster of Paris moulded around you. Glove dexterity wasn't very good, either.

"I had to go out the back to the adapter section of the spacecraft and one thing evident to me was where the spacecraft separated from the Titan booster was just like big jagged saw teeth about six inches deep – the kinda thing you wanted to stay away from.

"I had this bar to stand on – one foot under and one foot on top to just hold myself in place, and it's hard enough to do that when you are in street clothes – but try doing that in a pressure suit where you've got no feeling through the shoes. I got every valve open, every arm twisted and open and telescoped out. Just putting the harness on was an incredible task. All this time I was building up my heart rates and overpowered the environmental control system of the suit until it got to 100% humidity and fogged up my visor.

"I was ready with the AMU, ready to fly, ready for Tom to cut me loose. I had taken off the umbilical from the spacecraft and only had a nylon tether about 125 feet long. I was ready to go – except I couldn't see! I didn't know which way was up. I looked into the black night, and the Earth and the night looked the same. I couldn't see the horizon."

Cernan could feel the Sun scalding his backside where he had torn seven layers of insulation, and sweat poured out and stung his eyes. Unable to wipe it away, or wipe the fogged visor except to make a hole with the tip of his nose, Cernan then lost voice contact with Stafford. When they disconnected the umbilical cord to the spacecraft, communications were switched to a line-of-sight radio link that did not work from the back of the spacecraft. Stafford remembers, "This was a real tough moment – I had to get him back in because he couldn't see at all, and we lost one way of the two way communications – he could hear me, I couldn't hear him, just a squeak, so we worked out a kind of a code signal. Gene lost $10\frac{1}{2}$ lb in two hours and ten minutes of EVA. He was the first person to walk in space completely around the world."

Cernan said, "Tom and I laugh about it now, but, quite honestly, we had an unwritten rule that if I really had a problem out there and wasn't able to get back in, he would have to pull the umbilical out of the spacecraft, do his best to close the hatch, and come back alone. He knew it; and I knew it – we just didn't talk about it."

Stafford passed the bad news on to Houston through Carnarvon, "He's fogging real bad. It's far more difficult than it was in the simulations." Cernan tried increasing the flow of oxygen into his suit but that did not help and even the heat of the Sun did not clear the visor.

On orbit 32, Bill Garvin, Capcom at Carnarvon, passed up the message, "EVA – no go." The next short pass of 1 minute 40 seconds was not scheduled, but Houston unexpectedly decided to support it. At the Carnarvon station the canteen ran out of bread! A frantic call to town found a bakery able to supply enough bread at short notice and the staff were fed their breakfast of baked beans on toast.

Above the clouds over the Atlantic Ocean Cernan began the struggle of his life to get back into the spacecraft and lock the hatch shut. He had semi-closed his hatch during the spacewalk to stop the sun shining directly into the cockpit. Unable to see, he had to grope around to find the hatch. As Stafford reeled in the umbilical, Cernan was finding it impossible to squeeze his 183 cm body in the stiff, inflated spacesuit into the shape and size of the pilot's seat. With Stafford holding his boots down on the seat to stop him floating away, Cernan used the desperate strength of a doomed man to push and wriggle his body and suit into position.

As the boots crept over the front edge of the seat an agonising pain flooded his hips. He was still only half way in so he bent his head and tried to force the hatch closed but it only hit the back of his helmet. Stafford reached across and wrenched the locking

Figure 3.25. "Are you aware that in the time it takes for your beauty preparation, Eugene Cernan has walked around the world?!"

Figure 3.26. Gene Cernan (left) and Tom Stafford congratulate each other after their Gemini IX flight. On the carrier Cernan had to empty a litre of sweat out of his boots from his arduous spacewalk.

handle, squashing Cernan a bit further in and forcing the hatch down. Once the closing ratchet was engaged they were able to wind the hatch down, turn by turn, compressing Cernan immobile into the seat. Starved for oxygen, he was gasping with stress and pain as spots wheeled about before his eyes.

"Coming in – no problem," lied Stafford to Mission Control as he filled the cabin with oxygen. With the cabin pressurised the spacesuit became pliable and Cernan was able to move his legs and remove his helmet. Stafford took one look at Cernan's hot, flushed face and promptly sprayed some fresh water into it.

After the spacewalk the astronauts slept, and returned to Earth to be picked up by the carrier USS *Wasp*. Television cameras on the *Wasp* picked up the spacecraft as soon as it dropped under the clouds, only 3 kilometres away, and the world saw its first live splash-down. The Augmented Target Docking Adapter came down on 11 June 1966.

When Cernan took his spacesuit off on the carrier he emptied nearly a litre of sweat from the spacewalk out of his boots. "We came home and debriefed. We talked about it and said well, maybe working under water can partly simulate the zero gravity conditions. I went and did some work in a tank and I said, 'yeah, yeah – this is similar as long as you don't cheat by trying to swim,' and that started the simulations on Earth using water in tanks.

"We designed the footholds, the golden slippers, and a set of tasks for Buzz Aldrin to do in Gemini XII. After my flight, I told people, 'Anchor my feet, tie my feet down to that bar and I could do anything back there,' and it turned out to be true on Gemini XII. We designed this monkey box with all these tasks for Buzz and he went out there to prove what we learnt on Gemini IX – that we were right."

1966 **Gemini X**

Gemini X began the final phase of the programme and was the most complex mission to date, with a welcome return to spectacular successes. The taciturn pair of astronauts

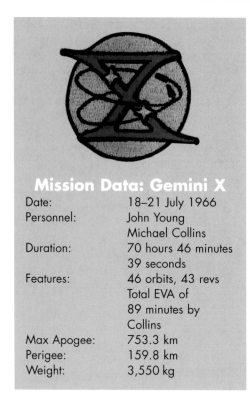

Mission Data: Gemini X

Date:	18–21 July 1966
Personnel:	John Young
	Michael Collins
Duration:	70 hours 46 minutes
	39 seconds
Features:	46 orbits, 43 revs
	Total EVA of
	89 minutes by
	Collins
Max Apogee:	753.3 km
Perigee:	159.8 km
Weight:	3,550 kg

sometimes tried Houston's patience to the limit. At one stage even Capcom No. 1 came on the voice channel, "John, this is Deke ... why don't you do a little more talking from here on?" The astronauts had been so busy with looking after the spacecraft, Agena commands, rendezvous procedures, navigation, as well as conducting experiments – they believed they were being asked to do four days' work in three – that there was little time for chatter.

They replied they had been talking, but to the Agena, "We have been pretty busy – this Agena takes a lot of talking to." When they met their own Agena on the fourth orbit, there was not a word from the astronauts down the radio channel, until Houston couldn't stand it any longer and asked through Tananarive, "Gemini X, are you there yet?"

"Roger, we're there," was the brief, illuminating reply from space.

Even Carnarvon found this crew unbearably quiet:

Carnarvon: "Gemini X, is there anything we can do for you?"

Gemini X: "No. Is there anything we can do for you?"

Carnarvon: "Yes – talk a little!"

The first rendezvous had used 64% of their fuel instead of the planned 40%, so they decided to use the Agena's motor and fuel as much as possible. The next day the Agena shoved the two docked spacecraft out to 764 kilometres above the Earth, higher than any man had been before, beating the previous record of 495.6 kilometres by Voskhod 2 on 18 March 1965. This high apogee was needed to place them in a position to rendezvous with the Apollo 8 Agena. As they were docked nose to nose with the astronauts facing backward they had a good view of the Agena rocket's 14 second firing over Hawaii. It was nearly sunset, so with the sun directly behind them, every spark and particle was brightly illuminated in front of the darkening background. Collins noticed a golden halo around the entire Agena.

Young recalls, "At first, the sensation I got was that there was a pop, then there was a big explosion and a clang. We were thrown forward in our seats. We had our shoulder harnesses fastened. Fire and sparks started coming out of the back end of that rocket. The light was something fierce, and the acceleration pretty good. Then the shutdown on the PPS (Prime Propulsion System, the Agena's rocket engine) was just unbelievable. It was a quick jolt ... and the tailoff ... I never saw anything like that before, sparks and fire and smoke and lights."

Unlike Conrad and Gordon later, they weren't bowled over with the panorama from the top, mainly because their view to the Earth was blocked by the Agena. They just kept an eye on the spacecraft systems and took some photographs.

Soon it was 1:00 am and time to sleep. Pulling out the metal window shades the astronauts were pleasantly surprised to find a sensual picture of a sexy girl pasted to each panel, looking quite out of place in their achromatic mechanical surroundings hurtling through empty space. As he settled down to sleep Collins found his arms were floating about in front of him and fretted they might drift out of control while he was asleep and inadvertently knock switches. Unable to find any way to keep them under control, dog-tired with the day's events, he fell asleep.

The next day they returned to their normal orbit height of 386 kilometres, 13 kilometres below the Agena 8's orbit, and 1,930 kilometres behind. The first spacewalk began in darkness as Collins took ultraviolet photographs of the moonless southern sky, followed by pictures in bright sunlight at different exposures of a titanium plate with red, yellow, blue, and green colour patches to see what effect space had on colour photography. In the middle of the colour test exposures both astronauts' eyes began stinging from lithium hydroxide accidentally released into the astronauts' air supply.

Young said, "Nobody ever found out what that really was. Our eyes started waterin' in the middle of the EVA and we couldn't see, so we had to stop the EVA. We went back on the main system and started breathin' again. It might have been something in the EVA oxygen system, but we never proved what it was."

The engineers suspected it was loose lithium hydroxide stirred up by using two air-circulating fans, and overcame the problem by only using one fan for the second space-walk.

The next day their Agena boosted them out to meet Armstrong and Scott's Agena. They had to rendezvous during the fifty five minutes of daylight so they could see their target, as by now it had no radar or lights. They found it flying along as steady as a rock, pointing straight up. Aware the astronauts were relying on ground tracking and their own eyeball navigation to rendezvous, Houston was standing by waiting for word of progress, but the voice channels were silent. Mission Control finally asked, "Gemini X, this is Houston, see anything of the Agena 8 around?"

"Yeah, we are about, I guess, seven or eight hundred feet out," Young told the surprised flight controllers. He then brought Gemini X about two metres under the Agena and Collins pushed himself off the open hatch towards the shining metal cylinder directly above. In seconds he bumped into the docking adaptor and managed to grip a channel of the docking cone and work himself around to the micrometeorite package. He swung onto it too fast, scrabbled at the shiny surface looking vainly for a handhold, and bounced off into space spinning out of control.

With the help of his nitrogen gun Collins returned back to the Gemini, and tried for the Agena again by propelling himself up to its nose. As he floated past he reached down into a recess behind the docking adaptor and grabbed hold of some wires to steady himself. This time he had no trouble releasing the square micrometeorite metal plate by pressing two buttons, but decided not to install a replacement as the Agena began to tumble with all his activity. Young called out a warning, "Come back, baby … get out of all that garbage."

Collins pulled himself back to the Gemini by his 15-metre umbilical, but discovered he had lost the Hasselblad SWC 70 mm still camera with all the spectacular shots of the spacewalk – the first camera to go into orbit. After another hatch opening to dump the unwanted gear overboard and another sleep period they returned to Earth on 21 July to be picked up by the USS Guadalcanal, while the Agena returned into the atmosphere on 29 December 1966.

1966 The Elation of Gemini XI

Around the middle of 1965 there had been talk of flying a Gemini spacecraft around the Moon in a mission called a LEO, or Large Earth Orbit. There had been sporadic interest from Congress down, but the top hierarchy of NASA felt it was best left to the Apollo missions designed for it. Conrad was very keen on the idea of going around the Moon, eventually persuading management to try a very high orbit in his Gemini XI mission instead.[32]

Gemini XI also tried something new – to meet a target in space in the first orbit. Conrad explained, "The big thing about this was there was no way we were going to get any help from the ground. Previously all the solutions and the phasing burns and all of that stuff was computed on the ground. We had to get ourselves into a matching orbit that was 15 miles (24 kilometres) smaller than the Agena target was in and phased the proper distance behind so that a while later we would begin the Terminal Phase Initiate (TPI) burn and go ahead and rendezvous with the target. The most

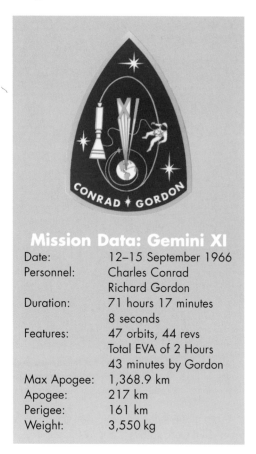

Mission Data: Gemini XI

Date:	12–15 September 1966
Personnel:	Charles Conrad
	Richard Gordon
Duration:	71 hours 17 minutes 8 seconds
Features:	47 orbits, 44 revs
	Total EVA of 2 Hours 43 minutes by Gordon
Max Apogee:	1,368.9 km
Apogee:	217 km
Perigee:	161 km
Weight:	3,550 kg

important thing was that we knew the exact time of our lift off down to quarters of a second. The ground had to pass up our corrected lift off time during powered flight just after launch, but before we disappeared over the horizon. We had this nice little hand-written chart which gave us the burn we had to do right smack at insertion.

"As we went over Madagascar the ground was going to try and pass up what they thought the TPI solution was, but we already had a good solution and caught up and rendezvoused shortly after Australia, and were flying some exercises around the target when we passed over Hawaii. There was a big dead period in the communications after Australia, so the ground were all very nervous waiting to find out what had happened when we reached Hawaii. The whole rendezvous was done either with our on board computer or the handy-dandy chart."

After liftoff, the astronauts steered Gemini XI to a safe dock over the Hawaiian Islands after only 80 minutes. "Mr Kraft – would you believe M equals one?" Conrad drawled with satisfaction, informing Houston they had successfully docked on the first orbit.

During the third day in the twenty-sixth orbit the Agena rocket belched fire to boost them to a new record height of 1,368 kilometres above the Earth, a height that clearly showed them the sphere of the Earth. "Whoop-de-doo, the biggest thrill of my life!" a gleeful Conrad called out as the acceleration shoved them into their straps – though they did wonder if the vehicle was ever going to come back as they blasted out into space!

Figure 3.27. "We're on top of the world," an excited Conrad called to Carnarvon during the 26th revolution of the Gemini XI mission, as they climbed to a record height of 1,368 kilometres over Brisbane, Queensland. The photograph shows the coast of north-western Australia, from Perth around to Darwin.

From 1,368 kilometres under the spacecraft Carnarvon called, "Hello up there!" and an excited Conrad burst out, "I tell you it's *go* up here, and the world is round … you can't believe it … we're on top of the world, we're looking straight down over Australia now. We have the whole southern part of the world at one window … utterly fantastic!"

Conrad remembers, "Australia was half in night on the ground and what we were seeing was the western coastline, there was a piece of beach there in the north west that was very prominent – Eighty Mile Beach, I think it was."

Figure 3.28. "We can't go on meeting like this, knowing there's a Yankee astronaut up there watching us!!"

Returning to around 290 kilometres they were supposed to get ready for their next space walk, but Conrad told Al Bean at Houston, "We're trying to grab a quick bite. We haven't had anything to eat yet today."

"Be our guest," offered Bean.

Over Madagascar, Gordon opened the hatch. "Here come the garbage bags …," said Conrad as everything in the spacecraft not fastened down floated out, including Gordon, before Conrad grabbed a strap on his leg. Gordon watched the sunset standing on the spacecraft floor, before photographing selected star fields. Then, deciding to keep the hatch open, the two astronauts simply fell asleep where they were! Conrad said, "We had worked three twenty-hour days; it got to be a nice quiet time in the day and we were waiting to get into a night pass." He called Houston after they woke up, "There we were … he was asleep hanging out the hatch on his tether, and I was asleep sitting inside the spacecraft!"

"That's a first," answered Capcom John Young. "First time sleepin' in a vacuum."

Gordon climbed out of the hatch and set up a 30-metre cable between the Gemini capsule and the Agena and they flew in formation. Instead of staying apart, the two vehicles tended to drift together.

"This tether is doing something I never thought it would do," reported Conrad. "It's like the Agena and I have a skip rope between us and its rotating and making a big loop. It's like we are skipping rope with this thing. Man, have we got a weird phenomenon going on here. This will take somebody a little time to figure out."

Conrad tried every trick he could think of to straighten the line. Although the line was curved, it seemed to still have tension. "I can't get it straight," he complained, but the ground engineers said to leave it alone. "So we really gritted our teeth and waited," Conrad said, and sure enough centrifugal force took over and the line smoothed out. They managed to use their thrusters to start the combination spinning once every nine minutes as they orbited the Earth. The cable remained taut and the two spacecraft happily spun their way around the Earth, while the astronauts then tired of watching the Agena and turned to eating their evening dinner.

Their rest was interrupted by the Hawaii Capcom suggesting they accelerate the spin rate. Although they had some initial problems with the line, "Oh, look at that slack! It's going to jerk this thing all to heck," called Gordon. It did stabilise again, and they were able to test their strange combination for artificial gravity. They put a camera against the instrument panel, and sure enough, when they let it go it drifted gently to the back of the cabin. The crew did not feel any physiological effects to themselves, though.[32]

Apart from problems with Gordon's spacewalk, the mission was a great success, and Gemini XI returned to earth under automatic control to be picked up by the USS *Guam*. The Agena came down on the 30 December 1966.

Conrad enjoyed this mission. "We got to fly the whole thing, which was the closest to the world I had left, that is flying airplanes. Like the M equals one rendezvous without help from the ground; we hand flew most of the burns – they weren't controlled by the computer; that sort of thing. It was a great flight."

1966 The Final Gemini Mission: Gemini XII

Lovell commented, "This mission was supposed to wind up the Gemini Program and catch all those items that were not caught in previous flights." Such as sorting out the Cernan and Gordon spacewalk problems. The Air Force wanted to try the Astronaut Manoeuvring Unit (AMU) from the Gemini IX mission again, but it was decided to keep Aldrin's spacewalk to conducting set tasks. Although Lovell was now a seasoned space traveller with the fourteen day mission behind him, it was Buzz Aldrin's first flight.

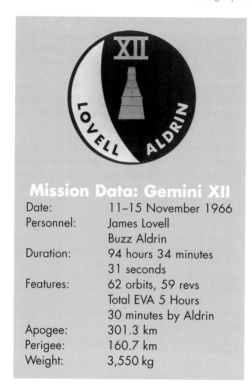

Mission Data: Gemini XII

Date:	11–15 November 1966
Personnel:	James Lovell
	Buzz Aldrin
Duration:	94 hours 34 minutes
	31 seconds
Features:	62 orbits, 59 revs
	Total EVA 5 Hours
	30 minutes by Aldrin
Apogee:	301.3 km
Perigee:	160.7 km
Weight:	3,550 kg

Unfortunately they weren't able to boost themselves up to the planned 740 kilometres high orbit due to a suspected faulty fuel pump in the Agena engine, so they were reassigned to witness a rare eclipse of the sun, west of the Galapagos islands, ending near Brazil. The two astronauts enjoyed a box seat view of the 8 second eclipse, 274 kilometres above the 800 scientists gathered below to watch the event in South America. Aldrin took excellent photographs of the eclipse, unaffected by the Earth's atmosphere.[32]

"We hit the eclipse right on the money, but we were unsuccessful in picking up the shadow," Lovell announced to Houston.

Using the experience and advice of Cernan in Gemini IX, Aldrin worked tirelessly training himself in a tank of water before the mission to work at all the experimental tasks until he felt he was perfect. During his long spacewalk Aldrin moved to a panel where he plugged in electrical cables, turned bolts, snapped hooks through rings, peeled off velcro strips, while experimenting with foot holds and his tether cable. He was able to complete all the tasks and suffered none of the fogging, perspiring, and tiredness of some of the earlier missions, although he did complain of cold feet.

Aldrin said, "In the first EVA I mounted a telescoping hand rail that went from one end of the spacecraft to the other; then I did some night photography, pretty much just standing up and shooting for ultraviolet light. In the longest EVA I had activities with the docked Agena. I used the handrail to move up hand over hand to the Agena at the

Figure 3.29. Lassoed! Conrad and Gordon spun around the world like a cartwheel, tied to their Agena rocket in this configuration to try artificial gravity. This picture is from the Gemini XII mission.

Figure 3.36. The Moon beckons: Gemini VII lines up the goal of Apollo in its sights.

4 The Apollo Project

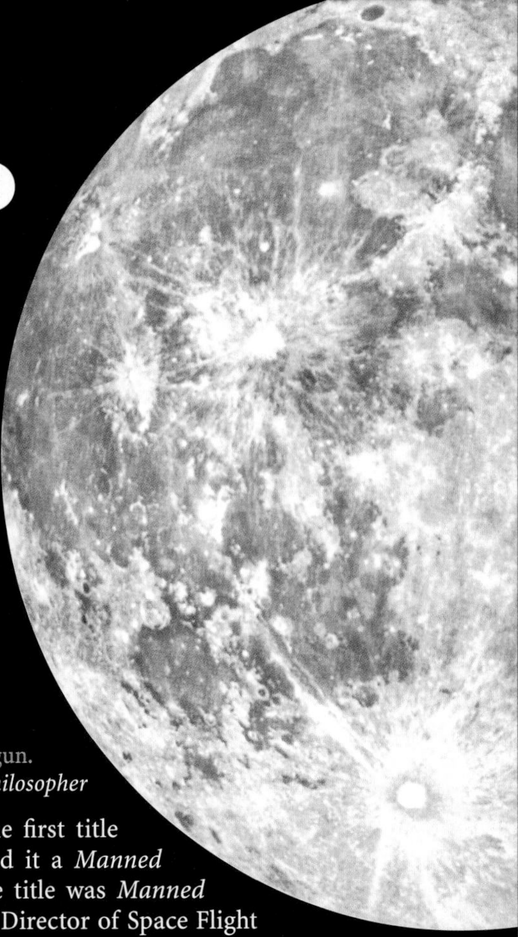

Earth bound history has ended. Universal history has begun.
Earl Hubbard American Philosopher

Project Apollo began under various names – the first title from a committee meeting in April 1959 called it a *Manned Lunar Landing Program*, and one cumbersome title was *Manned Lunar Landing Involving Rendezvous*. Abe Silverstein, Director of Space Flight Programs at NASA Headquarters, looked around for a programme name, but nothing came up that appealed, so in January 1960 he consulted a book of mythology. He said, "I thought the image of the God *Apollo* riding his chariot across the Sun gave the best representation of the grand scale of the proposed programme, so I chose it." Nobody objected and Hugh Dryden, the Deputy Administrator of NASA publicly announced it on 28 July 1960.[19]

NASA Administrator Webb said, "The Apollo Program grew out of a ferment of imaginative thought and public debate. Long range goals and priorities within our governmental, quasi-governmental, and private institutions were agreed on. Debate focused on such questions as which should come first – increasing scientific knowledge or using man–machine combinations to extend both our knowledge of science and lead to advances in engineering? Should we concentrate on purely scientific unmanned missions? Should such practical uses of space as weather observations and communication

◀ **Figure 4.1.** The view from the Moon: taken on Christmas Eve from Apollo 8 as humans witnessed the first earthrise over the lunar horizon. With north to the right of the picture, Africa is split by the sunset line and the continent of Australia is in darkness. Taken from 386,232 kilometres away from Earth, this is the picture that first awoke the people of Earth to the finiteness of our planet in the cosmos.

◀ **Figure 4.2 (inset).** The view from the Earth: Honeysuckle Creek shared Apollo 8's history making moments. The author took this picture of the antenna tracking Apollo 8 at about the same time as the astronauts took their famous picture of earthrise from the Moon (Figure 4.1).

relay stations have priority? Was it more vital to concentrate on increasing our military strength, or to engage in spectacular prestige building exploits?"[17]

The Goal of Apollo

Webb defined the goal of Apollo as: "To take off from a point on the surface of the Earth, travel to a body in space 240,000 miles distant, to go into orbit around this body, and to drop a specialised landing vehicle to its surface. There men were to make observations and measurements, collect specimens, leave instruments that would send back data on what was found, and then repeat much of the outward bound process to get back home. One such expedition would not do the job. NASA had to develop a reliable system capable of doing this time after time."[17]

Dr Robert Gilruth, Director of the Manned Spacecraft Center, explained, "Even before the President's decision to land on the Moon, we had been working on designs and guidelines for a manned circumlunar mission. This was done in a series of bull sessions on how we would design the spaceship for this purpose if the opportunity occurred. Our key people would get together evenings, weekends, or whenever we could to discuss such questions as crew size and other fundamental design factors. We believed that we would need three men on the trip to do all the work required, even before the complexity of the landing was added.

"The conceptual design of the moonship was done in two phases. The Command and Service Module evolved first as part of our circumlunar studies, and the lunar lander was added later after the mode decision was made."[17]

Figure 4.3. The huge Vehicle Assembly Building (VAB), 160-metres high, was the beginning of all the Apollo Moon voyages. Here all the component parts of the Saturn moonrocket were delivered from around the country and assembled under shelter from the weather. Without its protection even a 20 kilometre per hour wind could cause difficulties – but in Florida it also had to withstand hurricane force winds.

Figure 4.4. Moon Port on Merritt Island with the Apollo test vehicle on Pad 39-A. All the lunar missions, except Apollo 10, departed from this pad. Directly behind to the west is the VAB, 5.6 kilometres away. The Mobile Service Structure, parked half way on the left, is removed from the Saturn V about 11 hours before launch. Pad 39-B was 2.6 kilometres to the right, far enough away to be beyond the reach of a major explosion. In the optimism of the time there were plans for Pads C and D.

The Lunar Orbit Rendezvous Concept is Chosen for Apollo

At the embryonic stage the basic concept of the Apollo missions was to either blast a mammoth rocket directly to the Moon and return (Direct Ascent) favoured by some at NASA Headquarters, or assemble a large spacecraft in orbit around the Earth to land on the Moon and return (Earth Orbit Rendezvous), preferred by Brainerd Holmes and under consideration by von Braun. Nobody at NASA had looked seriously at Russian Kondratyuk's 1916 proposal of Lunar Orbit Rendezvous, breaking the spacecraft into smaller components at the Moon, which had also been read as a paper to the British Interplanetary Society by Harry Ross in 1948. A fourth possibility, never a serious consideration mainly due to the unknown capabilities of the navigation systems of the time, was Lunar Surface Rendezvous – sending a return spacecraft and consumables on ahead to land on the Moon before the manned flight.

In May 1960 a paper on the weight saving aspects of Lunar Orbit Rendezvous, with some brief calculations, was presented by William Michael of Theoretical Mechanics Division at Langley Research Center. Nothing came of the concept until in December 1960 John Houbolt, another engineer in the same division picked up the concept of

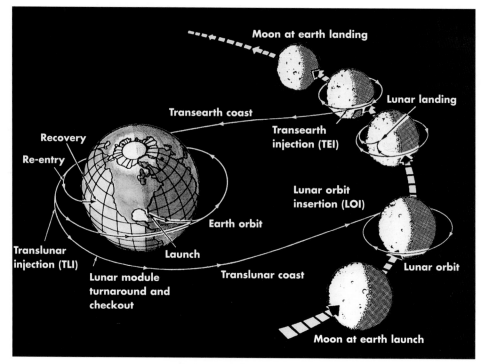

Figure 4.5. The Apollo plan.

Lunar Orbit Rendezvous and scribbled his own estimations on an envelope and gazed in disbelief at the figures. The weight of the launch vehicle would be cut in half!

With rising excitement he rushed off to his colleagues at the Space Task Group, but to his amazement they didn't believe him, or his calculations. Von Braun and President Kennedy's scientific adviser Wiesner all rejected the concept, assuring him it would not work. For more than a year Houbolt single-handedly hammered at the hierarchy of NASA, forming study groups and pushing the concept in all his lectures. Studies at the Manned Spacecraft Center, principally by Charles Frick, the Apollo Spacecraft Manager at the time, finally turned the tide and by the end of 1961 the Director, Bob Gilruth, and the Manned Spacecraft Center were converted. This acceptance initiated several months of computer analysis on a national scale; industry, universities, and laboratories were all briefed to examine every detail of the lunar rendezvous technique.

Gilruth visited von Braun at his headquarters in Huntsville, Alabama, on 5 July to present the results and his views. They all anxiously eyed von Braun. Studying the figures he remained silent for a while, then quietly admitted that the idea had merit. He was soon to put his whole weight behind the proposal.

Webb and Holmes at Headquarters were brought around and the united NASA team faced the last, and hostile hurdle – Wiesner, Kennedy's scientific adviser. Wiesner was unhappy about the crew performing intricate rendezvous techniques and orbital changes so far from Earth and help. Direct Ascent would not require rendezvous, and Earth Orbit Rendezvous would allow the chance of recovery of the spacecraft in an emergency.

During a visit to the Marshall Space Flight Center in Huntsville, Alabama, in September 1962 President Kennedy and a group were listening to an address by von Braun about the advantages of Lunar Orbit Rendezvous when Kennedy interrupted with, "I understand Wiesner doesn't agree with this?" The discussion almost became a brawl until Kennedy said it had yet to be decided, and later told Administrator Webb, "You're running NASA – you make the decision."

On 24 October 1962 Webb arrived at the White House armed with a barrow-load of independent studies and told Wiesner NASA was going ahead with building a Lunar Module unless the President himself objected. As Kennedy was awash with the problems of Khrushchev and the Cuban missile crisis nobody was game to distract him with a minor Moon landing problem, so the Lunar Orbit Rendezvous proposal became NASA policy.[17,24]

With tactful management and skilful leadership Brainerd Holmes, NASA's Head of Manned Spaceflight in Washington, organised a contract for Grumman Aerospace to build the Lunar Modules for Apollo within six months.

Dr Robert Seamans, Associate Administrator, recounted, "As planning for Apollo began, we identified more than 10,000 separate tasks that had to be accomplished to put a man on the Moon. Each task had its particular objectives, its manpower needs, its time schedule, and its complex inter-relationship with many other tasks. Which had to be done first? Which could be done concurrently? What were the critical sequences? How and where were major parts to be developed and made? How were they to be shipped?

"Apollo was an incredible mixture of large and small, of huge structures and minia-turised equipment. Throughout the programme we tried to maintain a flexible posture, keeping as many options open as possible. Alternate plans had to be quickly but carefully evaluated. The Apollo 8 flight was an example of the virtue of schedule flexibility. We could not take steps so small that the exposure to the ever-present risk outweighed minor gains expected. Yet neither could we take steps so big that we stretched equipment and people dangerously far beyond the capabilities that had already been demonstrated. We followed the fundamental policy of capitalising on success, always advancing on each mission as rapidly as good judgment dictated."[17]

By the end of 1967, America was launching the world's biggest booster, the mighty Saturn V rocket. The rocket was named Saturn because it is the next planet after Jupiter in our solar system and Jupiter was von Braun's earlier rocket. With the Saturn V Dr Werner von Braun reached his goal – his boyhood dream of building a rocket to send manned spacecraft into space was fulfilled.

The Saturn V Moon Rocket

It took 325,000 people to help von Braun produce the Saturn over a period of five years. In designing this big booster, the first consideration was for the safety of the astronauts. Second was weight. For every 0.45 kilograms of payload in a 4536-kilogram spacecraft, 34 kg of rocket thrust would be required. To add another 0.45 kg you were touching off a nightmarish chain reaction of more fuel, therefore more weight, therefore still more fuel. If von Braun added 45 kg to the Saturn's first stage he would be rid of it $2\frac{1}{2}$ minutes after liftoff. But if the flight engineers added a 45 kg to the lunar spacecraft, it had to be lugged around in space for up to 8 days.

The Saturn V, popularly known in the industry as "The Stack", was made up of three separate rocket motors, each with a special task. The bottom part, or first stage, was the most powerful working motor the world had ever known. Called the Saturn S-IC, it was built by the Boeing Company. Clustered around the base were five big Rocketdyne F-1 engines which pushed the 110.6 metre high, 3,000 tonne vehicle from rest on the pad to 9 times the speed of sound in the first 160 seconds of flight, consuming 2,128 tonnes of kerosene and liquid oxygen at a rate of 13.3 tonnes per second and producing an incred-ible 3.47 million kilograms of thrust!

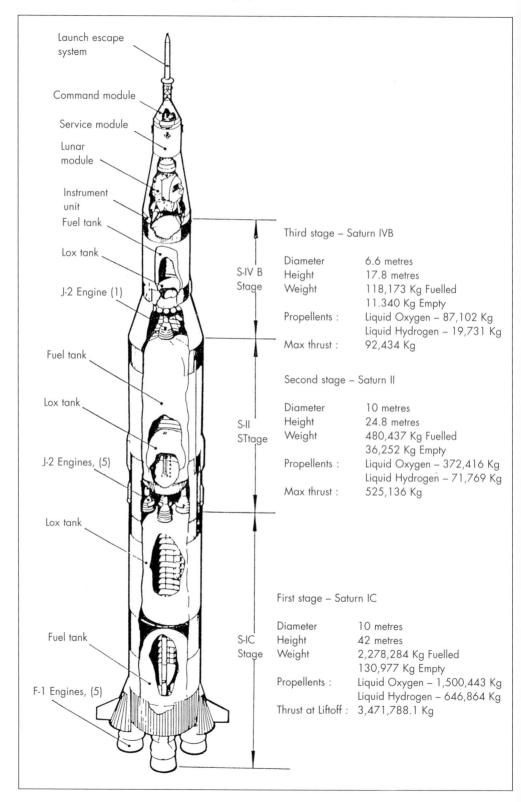

Launch escape system

Command module

Service module

Lunar module

Instrument unit

Fuel tank

Lox tank

J-2 Engine (1)

Fuel tank

Lox tank

J-2 Engines, (5)

Lox tank

Fuel tank

F-1 Engines, (5)

S-IV B Stage

S-II STtage

S-IC Stage

Third stage – Saturn IVB

Diameter	6.6 metres
Height	17.8 metres
Weight	118,173 Kg Fuelled
	11.340 Kg Empty
Propellents :	Liquid Oxygen – 87,102 Kg
	Liquid Hydrogen – 19,731 Kg
Max thrust :	92,434 Kg

Second stage – Saturn II

Diameter	10 metres
Height	24.8 metres
Weight	480,437 Kg Fuelled
	36,252 Kg Empty
Propellents :	Liquid Oxygen – 372,416 Kg
	Liquid Hydrogen – 71,769 Kg
Max thrust :	525,136 Kg

First stage – Saturn IC

Diameter	10 metres
Height	42 metres
Weight	2,278,284 Kg Fuelled
	130,977 Kg Empty
Propellents :	Liquid Oxygen – 1,500,443 Kg
	Liquid Hydrogen – 646,864 Kg
Thrust at Liftoff :	3,471,788.1 Kg

Stop and look at those figures again … and visualise a vehicle the height of a 36-story building, the weight of a navy destroyer, taking off and breaking through the sound barrier in 40 seconds, then reaching nine times the speed of sound in the next 90 seconds! An awesome spectacle of brute power and noise. The centre motor was fixed, but the four outer motors swivelled around to steer the vehicle. Resembling a shower rose the size of a manhole cover, the heart of the motor was the injector, which had to control the flow of 2.6 tonnes of fuel and oxidiser per second through 6,300 holes into the huge combustion chamber. It took 30 different designs over 18 months before they had a reliable injector. To test the cure for an instability problem that destroyed the motor during trials, they had to install a small bomb in the combustion chamber. The resulting explosion while the engine was running temporarily disrupted the flow of fuel to make sure the engine would recover and not destroy itself. Despite its vast size, a thumbprint left in the fuel tank could leave enough grease to react with the oxidiser to make an explosion.

The second stage was built by North American Aviation Rockwell International. Called the Saturn II, it was powered by five J-2 engines, burning liquid hydrogen and liquid oxygen totalling 523,324 kg of thrust, which pushed the vehicle to a height of 161 km before dropping off for the third and final stage to take over, a single J-2 stage of 94,459 kg thrust to shove the spacecraft into orbit from a height of 185 km. This third stage, built by McDonnell Douglas Astronautics and called the Saturn IVB, was used again to drive the Command and Service Module from 28,000 kilometres per hour in Earth orbit to 39,588 kilometres per hour to overcome the Earth's gravity and send the astronauts and their spacecraft out into space. The Saturn IVB accompanied the CSM to the Moon and in the later missions crashed onto the lunar surface for the seismometers to measure the effect of the impact.

Between these rocket motors and the manned spacecraft on the top was a circular collar called the Instrumentation Unit, or IU, which was built by IBM, and contained the controlling and guidance equipment for all the rocket motors.

Systems engineering and overall responsibility was given to the Marshall Space Flight Center, Huntsville, Alabama, under the direction of Wernher von Braun. A spokesman said, "Throughout Saturn V's operational life, its developers felt a relentless pressure to increase its payload capacity. At first, the continually growing weigh of the LM was the prime reason. Later, after the first successful lunar landing, the appetite for longer lunar stay times grew. Scientists wanted landing sites at higher lunar latitudes, and astronauts, like tourists everywhere, wanted a rental car at their destination. How well this growth demand was met is shown by a pair of numbers: the Saturn V that carried Apollo 8 to the Moon had a total payload above the IU (Instrumentation Unit) of less than 36,288 kg; in comparison, the Saturn that launched the last lunar mission had a payload of 52,618 kg."[17]

Why was the Saturn V so reliable? Von Braun explained, "It was not designed in the sense that everything was made needlessly strong and heavy. But great care was devoted to identifying the real environment in which each part was to work – and 'environment' included accelerations, vibrations, stresses, fatigue loads, pressures, temperatures, humidity, corrosion, and test cycles prior to launch. Test programmes were then conducted against somewhat more severe conditions than expected. A methodology was created to assess each part with a demonstrated reliability figure, such as 0.9999998. Total rocket reliability would then be the product of all these parts' reliabilities, and had to remain above a figure of 0.990, or 99%."[17]

◄ Figure 4.6. The Saturn V configured for the Apollo 11 mission.

On top of this monstrous fuel tank, sat two of the most complex and sophisticated craft ever built – the Command and Service Module and the Lunar Module, made up of four million parts. The super-concentrated development of various systems to support the Apollo missions produced many brilliant inventions that were lost in the immense project, but each playing a vital role in the ultimate success of the mission. The engineering accomplishments to achieve a lunar mission are beyond comprehension to anyone not directly involved. For instance, to the average citizen used to the normal picnic vacuum flask, it is hard to comprehend such revolutionary devices as the specially developed fuel tanks to preserve the super cooled fuel. If the ice from your refrigerator was put in this tank, and the tank was kept in your kitchen, it would take over 8 years for the ice to melt!

The Command Module was 3.4 metres long, 3.9 metres wide, and weighed 6 tonnes. It was the only part of the Stack that returned to Earth. As well as all the navigation, engineering, propulsion, and other mission requirements, it had to provide living accommodation for three people for the duration of the mission. It also had to be able to fly through the atmosphere, be a sophisticated spacecraft, mobile home, re-entry vehicle, parachutist, and boat.

The Service Module carried most of the consumables, power, and spacecraft environmental control system, for the journey to the Moon and back. It also had the SPS (Service Propulsion System), a rocket engine of 4,386 kg thrust for changing the spacecraft's course, such as going into and coming out of lunar orbit. The Service Module was a cylinder 7.3 metres long, 4 metres in diameter, weighing 26 tonnes fully loaded. The Command and Service Module, normally combined and called the CSM, were mated together for the whole journey until just before re-entering the Earth's atmosphere, when the Service Module was cast off, leaving only the Command Module, housing the astronauts, to return to the Earth's surface.

The combined Command and Service Module was usually referred to as the CSM. The picture of the Apollo 13 CSM being loaded onto the Saturn V (See Figure 6.7 on page 266) gives a clear view of what it looks like.

The Lunar Module, or LM, separated from the CSM in lunar orbit and landed on the surface of the Moon. Using its legs as a launching platform, known as the Descent Stage, the remaining part of the LM, known as the Ascent Stage, returned to the CSM, and was cast off in lunar orbit. In the case of Apollo 11, the LM was left in lunar orbit, but in the remaining Apollo lunar missions the LM was crashed onto the Moon's surface to provide seismic readings of the lunar interior on the instruments left behind by the astronauts. The whole LM stood 7 metres tall and weighed 16 Earth tonnes. Designed to work in only one-sixth gravity, no atmosphere, and with a critical launch weight, the LM was such a fragile craft that if a worker on Earth dropped a screwdriver it would easily punch a hole through the thin aluminium skin.

The computer on the LM was 61 cm long by 30.5 cm high, and 15 cm wide, and weighed 31.7 kg. Twenty-five years later it is hard for computer literate people to believe that the computer filled a box that size, with a memory of only 38K (38,916 bytes). The astronauts flew the LM standing up, cinched down with straps. It was very basic – there was nowhere to sit down and no lining on the insides. Weight was so crucial even the paint was specially selected.

Neil Armstrong, describing the Lunar Module, said, "The design was an evolutionary process. The cabin volume was changed from spherical to cylindrical; the landing legs were reduced from five to four; the window area was substantially reduced; and the seat removed. The standing pilot, or 'Trolley Conductor', cockpit reduced weight, moved the pilot's eyes closer to the windshield, improved visibility, and improved the structure. A

the service platforms. Babbitt was nearly over-whelmed by a sudden blast of pressure and searing heat with particles that speckled his clothes with burning holes. He heard voices calling, "She's going to blow. Clear the level!"

He turned to run, but recovered and collected three technicians with a shout, "Let's get that hatch off – we've got to get them out of there." They could only see a few inches

Figure 4.12. Astronauts Edward White II, Virgil Grissom and Roger Chaffee.

through the thick smoke that enveloped them, and had to make frequent dashes for the nearest fresh air. More helpers arrived. A few gas masks and fire extinguishers turned out to be quite useless, so the rescuers resorted to crawling around on the floor, trying to avoid the poisonous gases and smoke swirling around above them. In the blockhouse Deke Slayton, checking procedures with Capcom Stu Roosa, immediately ordered a medical team to the spacecraft.

It took five people six minutes to open the three layers of hatches and reveal the holocaust and meltdown within. The first rescuers to look inside could see little through the darkness and smoke. Carrying a torch, fireman James Burch pushed his way into the cockpit and first found White's feet and legs in the smouldering boots and legs of the spacesuit. He grabbed at a leg and was surprised to find it was still normal flesh – but the body was lifeless. He drew back and flashed his torch around; Chaffee was strapped in his couch, Grissom was sitting looking towards White who was lying on his back just below the hatch, his arms still reaching for the bolts. Some lights on the instrument panels were still glowing dully through the thick smoke.

Down below, the pad crew milled about, not knowing what had happened. Later 27 technicians were treated for smoke inhalation. The emergency escape procedures, to be tried later that day at the personal request of the astronauts themselves, planned to have the astronauts out of the spacecraft in 90 seconds – Grissom, White, and Chaffee died in less than 14 seconds.

The hatch that was supposed to protect them from the dangers of space, trapped them to a superheated death within. Officially, they died from suffocation by poisonous gases, mainly carbon monoxide.

Terry Kierans, M&O Supervisor at Carnarvon, remembers, "At first we thought the news coming down the communications line from Cape Canaveral was part of the simulation, until a teletype message explained the awful truth."

For a quick escape the Mercury spacecraft had an outward opening hatch. Gus Grissom nearly drowned in 1961 when the hatch of the MA-4 Mercury capsule inadvertently blew off and the spacecraft sank under him. This incident helped NASA decide to specify that the Apollo spacecraft should have an inward opening hatch. It is ironic that this newly designed hatch was sealed too tightly to be opened quickly after the fire had increased the internal pressure of the capsule. Twice Grissom was unfortunate enough to have the wrong hatch at the wrong time.

Figure 4.13. In the White Room of Pad 34 at the Kennedy Space Center is the blackened Apollo 1 Command Module that incinerated astronauts Grissom, White and Chaffee.

Due to the intensity of the fire caused by the pure oxygen under higher than normal pressure, and the destruction of the interior of the spacecraft the precise cause of the fire was never really determined. It started under Grissom's couch, and was apparently caused by a momentary arc in the wiring, possibly from damaged insulation. A sudden voltage fluctuation about nine seconds before the first cry tended to confirm this proposal. It took over seven hours to remove the bodies.

The shock spread across the world. The next day, NASA Administrator Webb said, "We've always known something like this would happen sooner or later … who would have thought the first tragedy would be on the ground?"

Typical of the fast-moving Webb, within three months he was getting the momentum of NASA going again. Vienna-born George Low, Deputy Director of the Manned Spacecraft Centre at Houston, was in a Grumman Gulfstream plane taxiing onto the runway at Washington airport, when it was recalled, and the passengers told to wait in the airport lounge room. In the lounge Low found: "Everybody in the line of command above me in NASA seemed to be there, everybody between me and the President. Jim Webb asked me to take over the management of Apollo. I probably would have liked some time to think about it, but since anyone I might have wanted to consult was already there in the room, there was no point in waiting. I said 'Yes, Sir'."

Low found the job overwhelming. "Dictating 18 hour days, seven days a week. My briefcase was my office, my suitcase my home, as I moved from Houston to Downey, to Bethpage, to Cape Canaveral, and back to Houston again. At Tranquillity Base, the Sun would only rise 33 more times before 1970."

During the modifications Low said, "Arguments got pretty hot as technical alternatives were explored. In the end I would usually decide, usually on the spot, always explaining my decision openly and in front of those who liked it least. To me this was the test of a true decision – to look straight into the eyes of the person most affected by it, knowing full well that months later on the morning of a flight, I would look into the eyes of the men whose lives would depend on that decision. One could not make any mistakes.

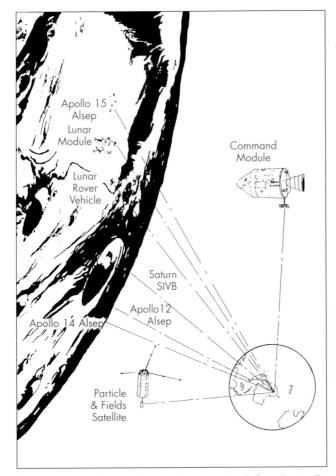

Command
Module

Apollo 15
Alsep
Lunar
Module

Lunar
Rover
Vehicle

Saturn
SIVB

Apollo12
Alsep

Apollo 14 Alsep

Particle
& Fields
Satellite

Figure 4.17. A diagram showing the many signals from the Apollo 15 mission to be collected at various times by the tracking stations for relaying back to Mission Control in Houston.

1966. Originally managed by the contractor, Standard Telephones & Cables Pty. Ltd. of Sydney, Honeysuckle Creek was officially opened by the Prime Minister, the Honourable Harold Holt, on 17 March 1967.

Simulations – Lunar Missions on Earth

Astronauts and ground teams around the world began knuckling down to tough simulations to sharpen their skills for all the lunar activity to come. While the astronauts and operational teams were testing their abilities, the spacecraft were undergoing rigorous tests too. Continuous testing, simulations, and developing procedures night and day was every Apollo worker's daily grind.

Robert Stanley, Head of the Mission Test

Figure 4.18. The Deep Space Station DSS14 64-metre Mars antenna on an azimuth/elevation mount. The single "S" Band feedcone was used to support the Apollo 11 and 12 missions, but by Apollo 13 it had a three-cone feed system. First operational in 1964, it was commandeered to support the Apollo programmes when required for the better signal quality provided by the bigger dish.

Figure 4.19. The Pioneer 26-metre Deep Space Station antenna (DSS 12) on an Hour-Angle/Declination mount, was built in 1958. Located about 6.4 kilometres north of the Apollo station, it was also commandeered to support the Apollo missions when required. It was closed down in 1981.

Management Organisation at the Goddard Space Flight Center in Maryland explains the organisation's function, "We were there to check out the ground tracking station's hardware, the software, the procedures and to train the personnel for each upcoming mission. We used an airborne system of three Constellation aircraft to carry sixteen racks of equipment that would generate all the Gemini and Apollo spacecraft signals, including Voice, Command, and Telemetry. It behaved just like a spacecraft except for the antenna pointing. We had an old World War II Norden bombsight in the cockpit as an aiming device so we could find the station and give the ground station predicts to point their antennas at us.

"Some of the trips we were away for three months – in the mid-sixties I travelled up to nine months of the year. I remember once they asked us to go up to Yokohama Bay in Tokyo to meet the tracking ship CSQ, the *Coastal Sentry Quebec*, for Gemini V and I was directing the simulations with the aeroplane and the ship was all torn apart while we were trying to train the guys. In walked Ed Fendell with this Flight team and Ed was all ticked off at me

Figure 4.20. The Receiver/Exciter system at Goldstone. Comprising gold-plated modules, it was responsible for sending and receiving the combined signals to, and from, the antenna.

Figure 4.21. Goldstone's Apollo 26-metre antenna on an X–Y mount, which allows the dish to track a spacecraft directly overhead with ease at a maximum speed of 3° per second. Built in 1966, it went on to support the Space Shuttle after Skylab.

because we were running the test and they couldn't put the ship back together to set sail for the mission."

The author remembers that the simulation teams and their aircraft would always arrive before a mission and for a week give the station and staff a thorough exam. No part of the station was exempt. "Once they came to me and I had to run to the store and get an obscure electronic component against a stopwatch. They would fail equipment, remove key personnel inferring a heart attack, and generally had a good time messing up the smooth flow of the pass procedures, concentrating on any new procedures. We always learnt a lot with each visit, and would go into each mission feeling confident we could handle all that was required of us.

Sometimes we got their own back on these simulation teams."

Figure 4.22. The Madrid tracking station in 1992. The Apollo 26-metre antenna is in the upper-left of the picture, and the 64-metre Deep Space antenna DSS63 (enlarged to 70 metres in 1987) is at the upper-right.

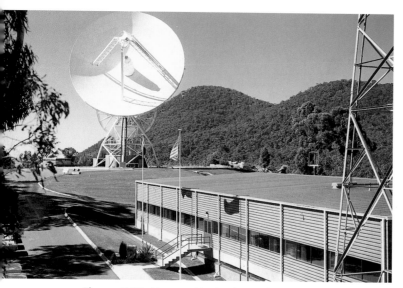

Figure 4.23. Honeysuckle Creek was the Australian Apollo station opened in March 1967. At the end of the Skylab programme in 1974, Honeysuckle Creek transferred to the Deep Space Network to track the Pioneer, Pioneer-Venus, Voyager and Helios missions, as well as the Viking landings on Mars. It was closed down in 1981 and the antenna relocated to Tidbinbilla.

Robert Burns, an Apollo Simulation team member from Goddard once experienced this, "An awful lot of funny things happened during simulations. The first one that comes to mind was a joke played on me at Honeysuckle. It was my first trip to Australia, I was rather new to this simulation business at the time – to say I was nervous and a bit edgy would probably be an understatement. I was the ground station Telemetry Observer and we had conducted several simulated spacecraft passes, and I was just beginning to feel I might know what I should be commenting on. I was listening to the various inhouse voice channels when I

Figure 4.24. The Unified "S" Band area with the Receiver/Exciter on the left, Ranging Subsystem at the back, and Antenna Positioning and Data Subsystems on the right. The Servo Console to control the antenna is in the foreground.

heard a sugary-sweet young lady's voice in my headset saying, 'Robby darling .... oh Robby daarling. ...'

"I broke into a cold sweat as the voice kept getting more personal by the second. I kept thinking that everyone could hear what was going on and I couldn't figure out how to get her to stop. I just wanted to crawl behind the cabinets and hide. It seemed to go on for hours, but it was probably only a few minutes. The pass was completed and the de-brief began. I was so flustered I not sure I made one intelligent statement. The Operations people assured me (so they claimed) I was the only one who could hear the comments."

Once, during the initial night testing of Honeysuckle Creek, Canberra residents flooded the switchboard with calls reporting strange lights and UFOs – nobody knew what this strange brilliant light was, ranging backwards and forwards in the sky to the south of the city. After one intense period of training a startled Honeysuckle Creek wife was woken up in the dead of night by her husband suddenly sitting up in bed, poking an imaginary button and calling out, "Decoms in lock!"

John Saxon, Operations Supervisor, recalls, "Later, as we got better, we used to involve the outside world

Figure 4.25. Bill Perrin at the Telemetry Monitoring Console with a complete picture of the telemetry data stream from the RF input to the final signal to line. On the upper right of the picture were the switching matrices used to select the data streams to the telemetry processing equipment.

Figure 4.26. Deep Space Station DSS 42 at Tidbinbilla which supported the Apollo missions as required. The additions for Apollo can be seen by the dark roof nearest the camera. The small dish on the tower is a link across the mountains to Honeysuckle Creek.

in our simulations, places like Sydney video. We actually managed to tie up most of the communications around the East coast of Australia when we had an internal simulation. Often Channel 7 didn't always get their news at the right time because we had all the television feeds tied up and they had to delay the evening news items."

1967 Apollo 4: The Mighty Apollo Saturn Rocket Enters Space

At exactly 7:00 am on 9 November 1967, one day before Surveyor VI landed on the Moon, the results of six years of intense work by von Braun and his team left the launch pad. Shock waves from the rocket shook trailers and rattled furniture in the brand new Launch Control Center at the Kennedy Space Center. Plaster dust covered everything. Nearly five kilome-

Mission Data: Apollo 4	
Date:	9 November 1967
Craft numbers:	AS-501/CSM-017
Personnel:	Unmanned
Duration:	8 hours 38 minutes
Features:	First Saturn V flight
	2 orbits
Apogee:	191 km
Perigee:	183 km

tres away the press site roof collapsed. The air pressure wave was detected at New York, 1770 kilometres away. The blast was compared to the explosion of the volcano Krakatoa. To reduce the noise of the rocket blast, sound suppression was built into subsequent Saturn V vehicles.

The resplendent Saturn V rocket heading for the heavens was the culmination of all the dreams of the space visionaries and pioneers throughout history. Apollo 4 was the first launch from the new Apollo Pad 39A, the first test of the Saturn V rocket, and the first trial of the integrity of the Command & Service Module and its heat shield. It put a record 126,529 kilograms into a 185 kilometre high orbit. The Command and Service Module was kicked out to 17,335 kilometres, and its rockets fired to hurl itself back into the atmosphere at 40,000 kilometres per hour, roughly the speed of a spacecraft returning from the Moon.

Landing only 9.5 km from the target, the scorched Command Module was picked up by the USS *Bennington* to show that everything had worked perfectly.

Before Apollo 4, NASA was doubtful about getting to the Moon before the end of the decade, Administrator Webb going so far as reluctantly announcing, "It's increasingly doubtful that an American – or Russian – will be on the Moon this decade."

Rocco Petrone, Apollo Program Manager from 1961–1966, said about Apollo 4, "To me that was the real mark; its success meant that we were really going to make the Moon landing."

After the success of Apollo 4 the goal to land a man on the Moon by the end of the decade was suddenly alive again.

1968 Next up for the Americans: Apollo 5

After weeks of bad weather and launch delays, on 22 January 1968 the 54.8 metre tall Saturn 1B that was to have launched Gus Grissom and his crew, soared into the Florida sky with the new Lunar Module, minus its legs, nestling in the nose behind protective panels. Apollo 5 was to try the Lunar Module's rocket motors in space four hours after entering orbit. The computer started the descent engine, but engineers at Mission Control watched in disbelief as it shut down after only 4 seconds. Luckily it turned out to be a computer programming error.

Then the ascent stage was boosted away from the landing frame, and again the engineers watched in horror as the ascent stage went into a wild dance. However it was another computer program error, and on the sixth orbit the Lunar Module returned into the atmosphere in a blazing fireball, with everybody satisfied that it was now ready for work.

Figure 4.27. The tracking station staff at Honeysuckle Greek occasionally had to risk danger when travelling the 52 kilometres to work, some of it through mountainous country. Once, the whole road slipped into the valley in the few minutes between two of the vehicles taking the shift members to work.

1968 The Untimely Death of Yuri Gagarin, First Man in Space

Officially banned from spaceflights, Yuri Gagarin had begun a series of refresher courses in conventional flying and decided he was due to log some more required flight time. He arranged to pilot a two-seater MIG-15 with

Figure 4.28. Sometimes, after heavy rain, large boulders would tumble down the steep mountainside to land on the road that led to the tracking station.

Mission Data: Apollo 5	
Date:	22 January 1968
Craft numbers:	AS-204/LM-1
Personnel:	Unmanned
Duration:	7 hours 50 minutes
Features:	Lunar Module only
	5 orbits
Apogee:	222 km
Perigee:	162 km
Weight:	42,506 kg

his Instructor Vladimir Seryogin. After spending some time with his daughters, he was joined at the airport by Seryogin and they left the runway in a pleasant spring morning. Some light frost still lay on the ground. Gagarin later radioed he was returning to base and requested clearance to land. The air traffic controller noticed from his radar that Gagarin was low and warned him to watch his altitude.

Nikolai Shalnov, a pensioner, was out walking his dog when he heard an

Figure 4.29. After one gruelling series of simulations, Honeysuckle Creek Station Director Tom Reid presented Team Leader George Harris from the Goddard Space Flight Center with a pair of sandals mounted on a plaque with the inscription:
"To Mr G. Harris, Jr in the hope that his feet will be dry when walking on the water.
Station Director and Staff, Honeysuckle." The station's kangaroo mascot helps to support the plaque.

aircraft close by, and as he looked up into the sky he saw a fighter plane plunge out of the storm clouds, skim some trees and explode onto the ground.

The first man in space, the ever cheerful Yuri Gagarin was dead at the age of 34, the cause suspected to be pilot error, although other reasons such as mid-air collisions have been put forward. Mourning its hero, Russian radio and television stations played long periods of gloomy funeral music, and the whole nation observed a minute of silence. The man who wanted to fly to Venus and Mars just missed seeing the flight of Apollo 11.

On 21 March Gagarin had been asked to sign a portrait of himself for a little girl's birthday. After writing his best wishes to Natasha, he asked the mother what date should he put on it.

"Oh, it doesn't matter – any date."

"When is her birthday?" asked Gagarin.

"The first of April," the mother replied, and Gagarin wrote 1 April 1968.

He died on 27 March 1968.

1968 **Apollo 6 Mission Still has Problems**

Apollo 6 put the Saturn V launch vehicle through its paces on its second launch to prepare it for manned flight, with further tests of the navigation systems and engines, and a final test of the Command Module's heat shield. Although there were some engine and fuel leaking problems, the spacecraft was successfully put into Earth orbit, and re-entered the atmosphere to land into the Pacific near the Hawaiian Islands to be picked up by the USS *Okinawa*. There were three major problems:

1. Two minutes and five seconds after launch the whole vehicle developed a

Mission Data: Apollo 6	
Date:	4 April 1968
Craft numbers:	AS-502/CSM-020
Personnel:	Unmanned
Duration:	9 hours 57 minutes
Features:	Second Saturn V test flight
Apogee:	364 km
Max Apogee:	22,258 km
Perigee:	180 km
Weight:	38,579 kg

Figure 4.30. Honeysuckle Creek's first Station Director, Brian Low, shows Dr George Mueller around the new tracking station in March 1967. His all up testing concept had men flying around the moon on the third flight, and landing on the sixth. Von Braun admitted Dr Mueller's concept was one of the main reasons NASA was able to meet President Kennedy's 1969 deadline. Dr Mueller was also involved in getting the Skylab missions under way using a surplus rocket casing for a vehicle.

lengthwise oscillation like a pogo stick, which, if severe enough, could destroy it.

2. A faulty manufacturing process caused the protective panels of the Lunar Module housing to fall off.

3. The most serious flaw was the unreliability of the second and third stage's J2 rocket motors to burn and re-start.

More than a thousand engineers and technicians ferreted out the causes of the problems from the miles of recorded data, as all the hardware was lost in space. Reliable solutions to all three problems were found.

The engine failures were a classic case of applied clever detective work on minimal clues. Under Paul Castenholz of Rocketdyne, the engineering team developed a reference pattern of the engine failures from the telemetry data and tried to duplicate failures to fit this pattern. They simulated every conceivable way to make these engines fail, but nothing would fit the reference pattern until they realised that the only thing they had not tried was the space environment, so they set up a test in a vacuum. Success – it failed! In two minutes eight liquid-hydrogen ASI fuel lines all failed at a bellows section. At last the pattern matched. All this testing took a whole month, but without computer analysis it would have taken a lot longer.[38]

1968 Apollo 8 Re-planned as Flight to the Moon. So Soon?

The Apollo 8 flight had been planned as an Earth orbit mission to check out both space-craft with a crew of McDivitt, Scott, and Schweickart, but the Lunar Module development was being left behind, with more delays expected. When the American CIA (Central Intelligence Agency) began issuing reports of the Russians recovering from the Komarov disaster and seemed to be planning a lunar flight with a new Soyuz spacecraft the Americans feared they were facing another space defeat, and decided something drastic had to be done.

Round about April 1968 George Low and Chris Kraft began toying with the idea of making the Apollo 9 mission into a circumlunar flight. Then in early July Slayton had to make a crew change to Apollo 9 when Mike Collins developed a bone spur in his neck, and he replaced Collins with Lovell. As McDivitt, Scott, and Schweickart had been working with their Lunar Module for eighteen months, Slayton moved them back to

Figure 4.31. An aerial view of the Manned Spacecraft Center in Houston, Texas, later to be called the Johnson Space Center. The Mission Control Center – Houston (MCC–H), the heart of all mission operations, was in Building 30 in the windowless left wing (circled). Inside were two nearly identical Mission Operations Control Rooms, (MOCRs) on floors 2 and 3. The second floor MOCR was used for Apollo 7, Skylab and Apollo–Soyuz missions; the third floor MOCR was used for all the Gemini and Apollo lunar missions, and, after Apollo 11, was designated a National Historic Site. The first floor was the Real-Time Computer Complex (RTCC) containing five main frame computers.

In the right wing were all the Staff Support Rooms (SSRs), housing the technical specialists responsible for supporting their counterparts in the MOCR.

During missions the Manned Space Flight Network (MSFN), the tracking stations, interfaced with the MOCRs through the Communications, Command, and Telemetry System (CCATS) on the first floor.

Apollo 9 with their Lunar Module, and moved the new crew of Borman, Lovell, and Anders forward to Apollo 8.

On 5 August 1968 the Spacecraft Program Manager, George Low, returned from a holiday and began to move on his proposal that Apollo 8 should go to the Moon without the Lunar Module. This would at least make sure of a lunar flight before the end of the decade, and also beat the Russians to the Moon.

He tried bouncing the idea off the NASA hierarchy. When General Sam Phillips presented the proposal to acting Administrator Paine in Webb's temporary absence, Paine reminded Phillips that the programme had fallen behind schedule, there had been seventy anomalies on the last flight, three engines had failed, they had not flown a manned Apollo mission yet, and there was a problem with the launch vehicle pogoing, a lengthwise oscillation like a pogo stick. "Are you really ready to go to the Moon so soon?" Paine queried Phillips. But he also felt that if all the weak spots had been identified, perhaps the proposal was feasible.

Webb, with Mueller at a conference in Vienna, was contacted by phone and predictably exploded on hearing the proposal, but Phillips and Low eventually brought him back to Earth. He agreed to think about the idea and it was generally decided to leave the final decision until after the Apollo 7 mission.[17] Webb, already suspecting he was in his last days as NASA Administrator, really hated the idea so soon after the Apollo 1 disaster, fearing another accident and another explanation to Congress. But after a technical briefing by Low and Phillips, Webb agreed the proposal should go ahead, but not to make it public.

Chris Kraft said, "I wasn't sure we could do it, but after looking at it for a few days we in flight operations were the ones that suggested we go into orbit, rather than just go circumlunar. When we did that the whole mission took on a different slant, of course, and I think that paved the way for the remainder of the Apollo program."

Figure 4.32. Christopher Columbus Kraft Jr: the original Flight Director, Director of Flight Operations, and Director of the Manned Spacecraft Center to become the Johnson Space Center. He was among the first 45 members of the Space Task Group formed in 1958 and became a legend in his own lifetime. He would say that the conductor of an orchestra does not have to know how to play all the instruments of an orchestra to make good music, similarly a good Flight Director can't know the intimate details of all positions under his control. His controllers disagreed with this philosophy – stories were legion to highlight his detailed knowledge of every mission and the minutia of every position in the Mission Control Center.

Slayton called McDivitt and told him the revised plans, then two days later he called Borman on a Sunday and told him a trip to the Moon on Apollo 8 was his if he wanted it. Without hesitating Borman answered, "Yes."

1968 **Russians Send First Lifeforms from Earth Around the Moon**

Now there was a definite date for NASA to send men to the Moon. But what were the Russians up to? On 14 September 1968 the Soviets sent their Zond 5 spacecraft off to the Moon with a D-1e booster on the first mission to round the Moon and return to Earth. Radio telescopes around the world tracked the spacecraft to try and determine the object of the mission. One report said that a tape recording of a voice reeling off simulated instrument readings was transmitted from near the Moon, which indicated that the Russians were testing voice communications from the Moon. Zond 5 rounded the Moon at a height of 1,950 km before dropping into the Indian Ocean on the night of 21 September, and its amazing cargo of turtles, meal worms, wine flies, lysogenic bacteria, chlorella, spiders' wort plants, and wheat, pine, and barley seeds were recovered for later analysis. Dr Vsevolod Antipov, a radiobiologist at the USSR Academy of Sciences, said that turtles were used because they had a sufficiently high level of complexity, but were easier to maintain than rats or mice. The two *Testudo horsefeldti* male turtles were each confined to their own plastic container, separated by a container of the other biological materials. Comparison with the control turtles that remained on Earth showed little significant effects of the space trip.[39]

Administrator Webb Hands NASA over to Dr Thomas Paine

In January 1968 President Johnson told Webb he was not going to seek re-election and Webb decided that on political grounds he should leave NASA, arranging to retire on his sixty-second birthday. He went to a meeting with Johnson at the White House on 15 September and Johnson immediately acted on a casual remark by Webb about retiring to announce the fact to the press. So a surprised Webb returned from the meeting no longer the Administrator. On 7 October 1968, just before Apollo 7 returned American astronauts to space, Webb addressed the Senate Space Committee with: "Last month I pointed out to the President that NASA has, with the recent induction of a number of experienced executives, the strongest and most capable organisation it has had in its ten year history. I told him it was fully capable of functioning at peak efficiency without me."

With his strong political connections and aggressive attitude, Webb had set a strong solid base for NASA to work from. Unfortunately, NASA never recovered the same momentum he generated with his clear-cut goals of landing on the Moon and exploring the solar system. His deputy since the beginning of the year, Dr Thomas O. Paine, took over the reins, and steered NASA through the Apollo 11 Moon landing mission.

Apollo Missions Carry Astronauts

1968 Apollo 7: American Astronauts Return to Space

The first three astronauts to ride a Saturn rocket into space were Wally Schirra, Donn Eisele and Walter Cunningham. Apollo 7 was also America's first manned

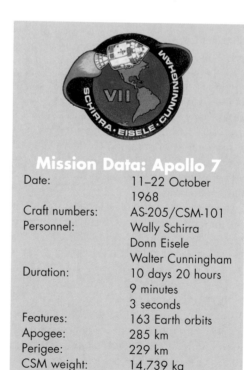

Mission Data: Apollo 7	
Date:	11–22 October 1968
Craft numbers:	AS-205/CSM-101
Personnel:	Wally Schirra
	Donn Eisele
	Walter Cunningham
Duration:	10 days 20 hours 9 minutes 3 seconds
Features:	163 Earth orbits
Apogee:	285 km
Perigee:	229 km
CSM weight:	14,739 kg
Distance travelled:	7,322,515 km

flight after Apollo 1's fatal fire. The main objectives of the mission were to check the new Apollo systems; to confirm that the Command and Service Modules were spaceworthy; to rendezvous with the Saturn IV-B booster; and to restart the Service Module's (SPS) rocket motor a number of times in space. The reliability of this motor was crucial to put the spacecraft in and out of lunar orbit in the missions to come.

Figure 4.33. Apollo 7 began the Apollo Program's manned flights. Donn Eisele (left) and Walter Cunningham (right) with their Commander, Mercury-veteran Wally Schirra, checked out the new Command and Service Module (CSM) for nearly 11 days in Earth orbit.

The only astronaut to fly in all three NASA programmes, Mercury, Gemini and Apollo, Schirra said before the flight, "We've had a goal that is a rather hard one to achieve, particularly one that we have had to follow when we lost three of our compatriots, and we don't want to make any mistakes that might cause something like that to happen again. We have not been the 'kid around' types that we might have been in the past; we're much more serious about it, because this is a much more complicated machine and there are many, many more people involved in it. I think you will find that you will see a good performance out of this total crew and we have tried very hard to make this machine work just the way it should."

One of Schirra's biggest concerns was the weather at launch. As it was the first time man was going to ride a Saturn rocket into space there was a chance they might have to abort the mission during the launch phase. As tests had shown that to land on hard ground with the early model Block 1 couches they were using would almost certainly injure the occupants, Schirra insisted that one of the launch rules would decree that it had to be an offshore breeze to a maximum speed of 18 knots for a safe launch so they would be sure of landing in the water if they had to abort.

On launch day, 11 October 1968, a high-pressure system over Nova Scotia created strong easterly surface winds over the launch area – blowing inland, straight towards the shore. At 11:02:45 EST the Saturn 1B rocket thundered off the pad, straight into a 20-knot easterly wind. From that moment Schirra was very angry – a potentially life-threatening rule had been broken right at the start of the mission.

He felt better as they progressed safely through the abort points, the Saturn rocket giving an easier ride than he had experienced in either the Mercury and Gemini flights. In less than ten and a half minutes they were in Earth orbit and were able to settle down to face a voyage of nearly eleven days circling the Earth.

To everyone's surprise it was on the ground that the unthinkable happened. Eighty minutes after Apollo 7 left the pad disaster struck at the very core of the brain controlling the mission – a power failure plunged the Mission Control Center in Houston into semi-darkness. The Flight Controllers stared helplessly at their black screens for two minutes, with only the dull glow of emergency lights around the room for their eyes to focus on. Outside the building, the mission happily pressed on until Houston recovered, and communications were restored.

During the first day of the mission Schirra developed the first head cold in space, soon infecting his companions. They found a common cold was much more uncomfortable in a weightless environment as the mucus accumulates in the nasal passages and does not drain out of the head. Blowing hard to remove the mucus only made the ear drums painful.

With his resignation from NASA already handed in, suffering with the head cold, and still angry at the irresponsible launch, Schirra took control of the mission and ran it his way. As one of the most efficient test pilots in the business he saw the flight primarily as an exercise to check out new mission procedures and a new spacecraft – a spacecraft that had just incinerated three of his friends. He refused to obey instructions from the ground if he judged it was not in best interests of the mission, or was not in the Mission Flight Plan. From his point of view television equipment and performing to a viewing public had no place in a technical test flight, but he had to accept the new NASA policy of providing the paying public with television broadcasts of the spaceflights. Thus Apollo 7 had the first American in-flight television, scheduled to be broadcast through the commercial networks after the rendezvous exercise with the Saturn IVB on the second day.

When Schirra woke up from a sleep on Saturday morning, he heard Mission Control requesting Eisele for a television transmission later in the day. As they had not checked the television system out, Schirra decided not to comply with this order, but Houston was insistent – Deke Slayton, the astronauts' chief, came on the link to the spacecraft:

Slayton: "Apollo 7, this is Capcom Number 1."

Schirra: "Roger."

Slayton: "All we have agreed to do on this is flip it. Apollo 7, all we have agreed to do on this particular pass is to flip, flip the switch on. No other activity associated with TV. I think we are still obligated to do that."

Schirra: "We do not have the equipment out, we have not had an opportunity to follow setting, we have not eaten at this point, I still have a cold, I refuse to foul up our timelines this way."

Figure 4.34. On the third day of the mission Eisele and Schirra held up a card asking their viewers to "Keep those cards and letters coming in folks" in the first live American television broadcast from space. The Wally, Walt and Donn Show was so popular they won a special Emmy Award.

Never had the spacecraft crew spoken to the ground controllers like this before. It was mutiny. From this point the mission became an uncomfortable confrontation between the ground flight controllers and the spacecraft crew, Schirra even taking on Deke Slayton and Chris Kraft. The mission rules had thought of every contingency except dealing with disobedient astronauts. In defence of this attitude Schirra later made the observation that none of the Flight Directors ever had to put their lives on the line for a mission.

Schirra found handling the much bigger Command and Service Module a very different proposition to the 'sporty' Gemini spacecraft. Without a rendezvous radar to provide the range and approach velocity, the crew had an anxious time judging the braking and distance to station keep alongside the Saturn IVB. Eyeing the tumbling rocket through his small window, Schirra felt it was a threat to their spacecraft, and kept at a safe distance of 30 metres.

It can become very boring enduring ten days in space with no entertainment to break the monotony of flight experiments and spacecraft housekeeping chores. With the Apollo 1 fire still looming over the project, no books or tapes were allowed on board. Tiring of the endless sunrises and sunsets every ninety minutes, the crew tried playing impromptu games such as throwing pens through holes made by circled fingers. They found they were never able to stop the pen tumbling enough to enter the hole.

The engineers at Houston insisted the crew follow mission rules and splashdown with their helmets on in case of a sudden loss of cabin pressure, but Schirra refused. The medical doctors were worried that with their colds the astronauts might suffer ear damage when they pressurised back to the Earth's atmosphere, so they told the astronauts to hold their noses, close their mouths and try to blow through their Eustachian tubes to keep the pressure in their middle ears in balance with the increasing air pressure. To do this they had to land without their helmets.

Splashdown occurred on 22 October, 370 kilometres south of Bermuda. There was a moment of anxiety when the capsule radio fell silent, but it was merely the rain squalls and choppy waves had capsized the spacecraft, putting the antennas under water. Helicopters from the carrier USS *Essex* picked the signals up as soon as the spacecraft flopped over right way up with the airbags.

All the mission objectives were successfully met. The rendezvous with the Saturn IVB was accomplished; the reliability of the Service Module Propulsion (SPS) motor was proved by starting it eight times, and as well as providing electricity, fuel cells supplied hot drinking water for the coffee Schirra had insisted on taking. For the first time in American missions a live television programme was broadcast and a mixed cabin atmosphere of 65% oxygen to 35% nitrogen was tried.

NASA was pleased and relieved with the results. If Apollo 7 failed badly a Moon landing before the end of the decade would have been doubtful. Apollo Program Director General Sam Phillips announced the result of the mission, "We achieved 101% of our objectives."

Flight Director Gene Kranz added, "This mission was a spectacularly successful test of the manned maiden voyage of the Apollo Command and Service Module, however learning to work with Wally in flight, well, in my book I call him 'The Grumpy Commander'. Unfortunately, he got this cold in flight, and space is not too tolerant to people who end up with a lot of congestion. It's very difficult to sleep and continue working – I guess he had his hands full. I never quite figured out why Wally had such a burr under his saddle there."

Director of Flight Operations Chris Kraft said, "Schirra was recalcitrant – the whole crew was recalcitrant as a result, and I imagine there was some justification for that on the basis of what happened on the pad when we killed three people. As fliers they performed the flight well, they just gave us a hard time on the ground." Kraft never spoke to Schirra about the crew's belligerent attitude to the ground, and Cunningham and Eisele never flew in space again.

1968 **Russians Try a Rendezvous between two Spacecraft**

The Russians were still busy trying to beat the Americans to the Moon. On 26 October 1968 they launched Soyuz 3 with Georgi Beregovoi aboard. Forty-seven-year-old

Beregovoi became the oldest astronaut into space at the time, after he flew around the Earth for four days. Ground controllers brought the unmanned Soyuz 2, launched the previous day, within 200 metres of Soyuz 3, when Beregovoi took over manual control to dock, but ran into problems when only 1 metre apart, and had to give up the docking exercise.

In November an unmanned Zond 6 flew around the Moon, but the cabin depressurised on the way back, and the capsule slammed into the Earth at a speed that would have killed a human crew.

A Zond 7 flight was planned with two cosmonauts for early December, just ahead of the Apollo 8 launch. But, still bogged down with a swag of technical problems, the Russians saw reason and scrubbed the mission two days before the scheduled launch, and the cosmonauts had to watch helplessly from Earth as Apollo 8 headed for the Moon to be the first humans to leave the vicinity of the Earth. When an unmanned Zond 7 was finally launched in January 1969, the booster rocket exploded and the Russians were in trouble again.

Some notable scientists of the time were still arguing against sending men to the Moon, regarding it as wasteful and dangerous. Machines can do the same job, they said. Speaking for NASA at a press conference at the Kennedy Space Center, Apollo 8 Astronaut Anders defended the plan to send men to the Moon. "Although Apollo 8 is not primarily a scientific flight, it would give science its first chance to have in the vicinity of the Moon an eyeball connected to a brain connected to an arm that can write or a tongue that can speak."

The Russians agreed. Mstislav Keldysh, head of the Soviet Academy of Sciences, said, "Man will always strive to take a direct path in scientific space research. Automatic devices can never fully replace man."

1968: November NASA is Ready for the Moon

By 11 November all the ground work for a lunar mission for Apollo 8 was complete and the message officially advising the President was laid on Lyndon Johnson's desk as he was handing over the Presidency to Richard Nixon.[17]

Lovell and Borman were only too happy to go off to the Moon – after living in Earth orbit with each other for a record 14 days in the claustrophobic Gemini VII, the thought of repeating it again had no appeal at all.

Excitement mounted when NASA's new Acting Administrator Paine announced on 12 November: "After careful and thorough examination of all the systems and risks involved we have concluded that we are now ready to fly the most advanced mission for our Apollo 8 launch in December, the orbit around the Moon."

Paine's announcement was backed up by President Johnson with a message to the astronauts saying: "I am confident that the world's finest equipment will strive to match the courage of our astronauts. If it does that, a successful mission is assured."

1968 Apollo 8: Space Travellers Leave the Earth for the First Time

The first of the manned lunar missions, Apollo 8, was regarded by many as being as significant, if not more significant than Apollo 11, when everything had been done except an actual landing. Director of Flight Operations Chris Kraft commented, "From Apollo 8 we really knew what we were doing. It was the boldest decision we made in the whole space programme – period!"

Flight Director Glynn Lunney said, "I thought Apollo 8 was the decision which opened the gate and let us slide down the hill to the landing. We knew we could go to the Moon,

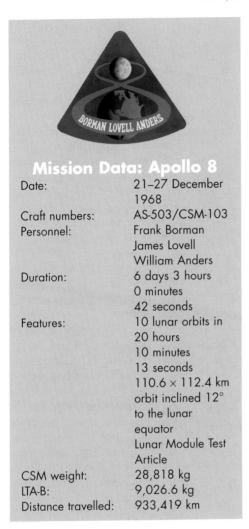

Mission Data: Apollo 8

Date:	21–27 December 1968
Craft numbers:	AS-503/CSM-103
Personnel:	Frank Borman
	James Lovell
	William Anders
Duration:	6 days 3 hours 0 minutes 42 seconds
Features:	10 lunar orbits in 20 hours 10 minutes 13 seconds
	110.6 × 112.4 km orbit inclined 12° to the lunar equator
	Lunar Module Test Article
CSM weight:	28,818 kg
LTA-B:	9,026.6 kg
Distance travelled:	933,419 km

go into and come out of an orbit around the Moon, do all the navigating, and get the Command Module back home."

Flight Dynamics Officer Jerry Bostick added, "From a trajectory viewpoint, it meant we had to accelerate some of the software in the Mission Control Center and the spacecraft and the worldwide tracking network. Now management had decided to go into lunar orbit it required very accurate calculations. I have told people that shooting for the Moon is a bit like duck hunting – you don't shoot at the duck, you shoot at a spot in front of it and let it fly into it. So you have to aim at a spot in front of the Moon equivalent to the thickness of a sheet of paper when viewed from Earth. We had confidence in being able to do this, but were a little nervous about doing it for the first time and much earlier than planned."

The author adds, "I remember that at the start of the mission I was feeling very excited but more apprehensive about this mission than any other, including Apollo 11, because it was the first time we were going to actually leave the Earth, away from being able to return quickly to safety in an emergency. I felt quite confident about Apollo 11 because Apollos 8 and 10 had been so successful – everything seemed to be working."

There were so many firsts for Apollo 8. The spotlight swung to the three big 26 metre tracking stations as it was the first time the larger antennas were really needed for manned spaceflight. It was the first time astronauts had experienced the full 3.4 million kilogram thrust of the big Saturn V; the first time man had left the planet to head into space; the first time anyone saw the whole Earth suspended in space; or could cover the whole Earth with their hand; the first time humans had not experienced a night with sunrises and sunsets; the first time to orbit the moon; the first time men had been occulted; first to see the backside of the moon; first to see Earth rise at the Moon, and first re-entry from the Moon.

Never had anybody travelled so far and so fast. Never had humans been so remotely isolated from their own kind. As with everything else in Apollo, the navigational accuracy required for a successful mission was mind-boggling. As with everything else in Apollo, the layperson could merely marvel and trust in the outstanding abilities of the technocrats. Borman was awed by the accuracy of the calculations of the planners when he was told from the flight plan to look down at a specified point in time, and to the second saw the sunrise strike a nominated feature on the lunar surface.

For a lunar flight the navigation had to take into account all the moving bodies involved – an Earth travelling at 1,609 km as it rotated, the spacecraft leaving the Earth at 40,232 kilometres per hour, travelling to the Moon 376,737 km away, taking into account the effect of the interacting gravity fields along the way, and arrive exactly 128.7 km ahead of the Moon, which itself is travelling at 3,219 kilometres per hour relative to the Earth. An error of only 1.6 kilometres per hour in the spacecraft's speed would mean missing the Moon by 1,600 kilometres. One NASA engineer was heard to say, "It was like running across the front of a speeding locomotive close enough to get a paint sample off its front without getting hurt."

Figure 4.35. Go slow to go fast. The journey to the Moon begins at a snail's pace as Apollo 8 crawls to the launch pad and history at 1.6 kilometres per hour. Weighing 9,150 tonnes with its load, the crawler steadied the "Stack" to within 1° of absolute vertical, even going up the 5° incline to the launching platform. Forty seconds after leaving the launch pad, the 3,000 tonne Saturn V was breaking the sound barrier!

Apollo 8 also gave America the race to the Moon – with Apollo 8 a success the Russians had lost any chance of sending the first humans to the Moon.

It took 8 years to get Apollo 8 ready for launching. Parts came from all over the country – the first stage came from Louisiana, the second and third stages from California, after testing in Alabama. Guidance and navigation equipment came from Wisconsin to be checked in Massachusetts, while the internal systems came from Florida and New Hampshire. All these disparate parts had to fit flawlessly together for a successful mission, a remarkable achievement in the time available.

The weather at Cape Canaveral on launch day was fine with some cirrus clouds high in the sky, while on the ground a 20 kilometres per hour breeze from the north kept the temperature a cool 15°C. At Goldstone over on the other side of the continent the key tracking station for the lunar departure was welcoming the first of the 12 hour shift arrivals at 3.00 am local time. They had travelled for an hour from the nearby town of

Figure 4.36. 21 December 1968 – Apollo 8: Dawn of a new age. ▶

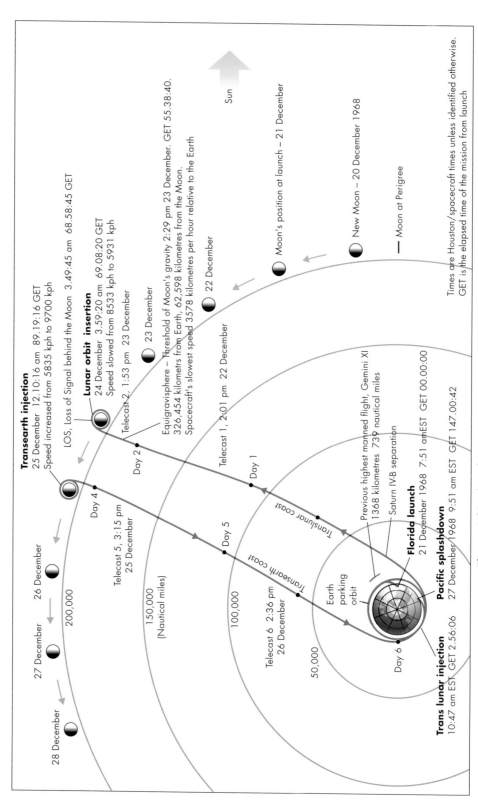

Transearth injection
25 December 12:10:16 am 89.19:16 GET
Speed increased from 5835 kph to 9700 kph

LOS, Loss of Signal behind the Moon 3.49:45 am 68.58:45 GET

Lunar orbit insertion
24 December 3.59:20 am 69.08:20 GET
Speed slowed from 8533 kph to 5931 kph

Telecast 2, 1:53 pm 23 December

Equigravisphere – Threshold of Moon's gravity 2:29 pm 23 December. GET 55.38:40. 326,454 kilometres from Earth, 62,598 kilometres from the Moon. Spacecraft's slowest speed 3578 kilometres per hour relative to the Earth

Telecast 1, 2:01 pm 22 December

Day 2

23 December

22 December

Moon's position at launch – 21 December

New Moon – 20 December 1968

—— Moon at Perigree

Sun

27 December   26 December

28 December

200,000

Telecast 5, 3:15 pm
25 December

Day 4

150,000
(Nautical miles)

100,000

Day 1

Day 5

Transearth coast

Transeach coast

Transthat coast

Transhar coast

Earth parking orbit

Florida launch
21 December 1968 7:51 am EST GET 00.00:00

Previous highest manned flight, Gemini XI
1368 kilometres 739 nautical miles

Saturn IV-B separation

Pacific splashdown
27 December 1968 9:51 am EST GET 147.00:42

50,000

Telecast 6 2:36 pm
26 December

Day 6

Trans lunar injection
10:47 am EST GET 2.56:06

Times are Houston/spacecraft times unless identified otherwise.
GET is the elapsed time of the mission from launch

Figure 4.37. Plan of the track to the Moon followed by Apollo 8.

Early on the morning of 24 December in the Mission Control Center all the flight controllers on duty suddenly sat up in surprise. For the first time the big centre screen switched over to a picture of craters and mares showing names such as Grimaldi, Gilbert, and Mare Crisium. For the past three and a half years the screen had only showed a picture of the Earth with a moving dot to indicate the spacecraft's position.

It was an exciting moment. At last they felt the mission really had reached the Moon, and all the readings from Apollo 8 showed a healthy spacecraft.

From across the void, Jerry Carr at the console in Houston announced the decision. "This is Houston at 68 hours and 4 minutes. You are *go* for LOI (Lunar Orbit Insertion)."

Borman replied, "OK. Apollo 8 is *go*."

"You are riding the best bird we can find," said Carr, cheerfully tried to hide the anxiety in the air. "We'll see you on the other side. One minute until LOS (Loss Of Signal behind the Moon). All systems are *go*. Safe journey, guys."

"Thanks a lot, troops," Anders responded.

"We'll see you on the other side," added Lovell.

At 3:59 am on Christmas Eve Apollo 8 slipped behind the Moon's rim and the communication channels fell silent. The predicted timing was so accurate Borman mused, "That was great, wasn't it? I wonder if they turned the transmitter off?"

While an expectant Earth waited, the three astronauts felt suspended in a black void, travelling upside down and backwards, only aware that the black surface of the night side of the Moon beside them was blotting out the stars.

Lovell thought, "I should be feeling something, something profound, but there seemed to be little to be profound about. There was nothing to indicate this monumental event is taking place. We were weightless moments before; we are weightless now. There was blackness out the window before; there is blackness now."

The three astronauts looked at each other in silence. Without the comforting voices from Houston to confide in and tell them what was going on, they were committing themselves to lunar orbit on their own.

Borman broke the silence, "So, are we *go* for this thing?"

"We're *go* as far as I'm concerned," Lovell checked his instruments.

"*Go* on this side," said Anders.

Lovell typed the instructions into his computer and watched for its response. "99:40" appeared in the readout – the code to say everything was *go* and the computer was waiting for the pilot's final approval. Lovell took a breath and pushed the "Proceed" button.

Borman had already aligned the spacecraft for the burn hours before. The SPS engine fired for the specified four and a half minutes and the astronauts felt a gentle pressure on their backs as the spacecraft slowed. With no atmosphere outside the astronauts could only hear a slight hum and felt the space-

Figure 4.39. The Apollo 8 crew: Mission Commander Frank Borman (left), Command Module Pilot James Lovell (centre) and William Anders, filling in the Lunar Module Pilot's position.

craft vibrate to the rocket's firing. "Longest four minutes I ever spent," murmured Lovell. Borman's heartbeat jumped to 130 beats per minute. If the rocket motor's first burn to put the spacecraft into lunar orbit had misfired they would have crashed onto the Moon's surface, or skipped off into an endless solar orbit.

Lovell looked at Borman and stuck his thumbs in the air. As Borman returned a thoughtful smile, Lovell turned back to his readouts. The Delta V (change in spacecraft velocity) reading showed 2,800 feet per second and as he watched the orbital parameters appeared: 60.5 and 169.1 miles. Just about perfect figures.

They were in an elliptical lunar orbit, with a closest approach of 97.4 kilometres rising to a maximum height of 272.1 kilometres.

"We did it," a jubilant Lovell called out.

"Right down the pike," agreed Anders.

"Orbit attained," the cool Commander announced. "Now let's hope it fires tomorrow to take us home again."

Up to now they had not seen the lunar surface so Borman reached for the spacecraft controls and brought it around 190°. He was the first to see the moonscape below roll into his window, then Lovell gazed in astonishment as an amazing panorama of sunlit tips of mountains, followed by tops of craters and mounds poked up through the blackness below the spacecraft.

"Oh, my God ...," Anders exclaimed, spellbound.

"What's wrong?" Borman swung around, prepared for an emergency.

"Just *look* at that!" Anders couldn't take his eyes off the scene as the sunlight spread over crevasses, craters, rilles and barren mountains that appeared in their windows as the spacecraft crossed the terminator, the division between night and day. It spread away in all directions to the lunar horizon, an alien, hostile land that no human had ever seen before, breathtakingly clear and stark with no atmosphere to blur the details. As they sped around the back they entered a moonscape bathed in brilliant white sunlight, covered with craters and sterile mountains. The craters were endless, of all sizes, and everywhere. For a moment the three astronauts forgot the technical world of Apollo and Mission Control as they pressed their faces to the windows. Lovell feels seeing the far side of the Moon was probably the high point of his long space career.

Figure 4.40. Life for staff on a tracking station varied from changing a transistor to changing a feedcone, such as Honeysuckle Creek had to in the middle of the Apollo 8 mission.

The author remembers, "Honeysuckle Creek was tracking the trailing edge of the Moon, pinpointing the spot where Apollo 8 should emerge. I was there and for once we could see the target on the antenna television boresite monitor. The Moon showed up clearly on the screen, the cross-hair aimed right on the edge of the crescent – steady as a rock – waiting. Right above the screen a large digital clock silently flicked the seconds aside. These weren't ordinary seconds; these seconds were counting down to a monumental success or a monumental failure of the mission. With all my equipment ready, I went into a anticipatory trance, mesmerised by the clock. As Chris Kraft said to me later, "If you weren't shaking at that point you didn't understand the problems".

The receiver operators were anxiously clamped to their controls watching the noise from the receivers, spring loaded to grab the first signs of a signal on their displays.

Suddenly there it was – the receivers locked up, and signals flooded through the station equipment, filling all the meters and dials with meaningful figures. Houston received the output from Honeysuckle Creek and a jubilant public affairs announcement said, "We've acquired a signal but no voice contact yet. We are looking at engine data and it looks good. We got it! We've got it! Apollo 8 is in lunar orbit."

Alan Foster, on the receivers at Honeysuckle, tells us, "I just made a normal acquisition as they came over the lunar horizon – it was a good signal, clean and sharp, no fading at all, one of the easiest acquisitions I had ever done because there was no antenna searching around, we could see the crescent Moon. I was relieved I can tell you. I have always remembered Network saying on the loop, 'That was a beautiful acquisition, Honeysuckle'."

Operations Supervisor Saxon said, "Because we weren't tracking the Lunar Module we had all our equipment configured onto the Command Module. There was a planned mode and an unplanned, or back-up mode. We decided we should cover the back up mode just as carefully as the prime mode. Unfortunately there were so many ways to configure this – we ended up locking up all the telemetry and sending it back to Houston all right – but there was a degree of confusion about where the voice was actually coming from. I remember hearing the public affairs loop saying we have data but no voice, and here I was frantically pushing buttons trying to find where they had put this voice, and Kevin Gallegos at the demodulators was pushing buttons as well, and no doubt Houston was also pushing buttons. I had my fingers poised ready to call the astronauts and tell them that we did have communications with Houston, but somehow we can't get you through. We managed to sort it out in the end – I was very close to being the first guy to speak to someone in Lunar orbit!"

Jerry Carr at Houston called out, "Apollo 8. ... Apollo 8. ..."

Lovell replied, "Go ahead, Houston. This is Apollo 8. Burn complete. Our orbit is 169.1 by 60.5 miles."

Carr said with relief, "Roger, Apollo 8. 169.1 by 60.5. Good to hear your voice."

Lovell's prosaic voice came down the channels from the Moon, through Honeysuckle Creek, and flashed over the Pacific to Houston. In the flight operations room at Houston the relief brought on wild cheering, prompting the duty Flight Director Glynn Lunney to say, "I'm sure it can be described as one of the happiest Christmas Eves just about anyone there had seen."

Impressed with the accuracy of Houston's prediction of the signal being cut off by the edge of the Moon as they went behind it they checked:

Anders: "Houston, for your information, we lost radio contact at the exact second you predicted."

Houston: "Roger, we concur."

Anders; "Are you sure you didn't turn off the transmitters at that time?"

Houston: "Honest Injun', we didn't."

While Borman busied himself with the status of the spacecraft Anders and Lovell were enthusiastically reporting their observations back to Earth. Lovell commented on his

Figure 4.41. On the fourth day, astronauts gathered around Deke Slayton (seated centre) to study a mosaic map of the lunar surface passing under Apollo 8. Standing (left) Harrison Schmitt with Buzz Aldrin. Gerald Carr (seated left) looks on with an engrossed Neil Armstrong. This scene was only days away from Armstrong and Aldrin being told by Slayton they were to join Mike Collins for the Apollo 11 mission to land on the Moon.

impression of the Moon's other side, "The backside looks like a sand pile my kids have been playing in for a long time. It's all beat up with no definition. Just a lot of bumps and holes."

Anders view of the back was: "It looks like a whitish gray, like dirty beach sand with lots of footprints in it. Some of these craters look like pickaxes striking concrete, creating a lot of fine haze dust."

Borman added, "To me it looked more like the burned-out ashes of a barbecue."

Later Borman cut into the chatter with a request for a SPS motor status report and, "We want a *go* for every rev. please, otherwise we'll burn in TEI one at your direction." He was warning everyone to keep on the ball or they would come home after the first orbit.

A later burn put them in a circular 96.5 kilometre orbit. As Apollo 8 circled the Moon they took hundreds of photographs, endless measurements and observations, and related their views back to Houston. Anders was mainly responsible for the photography, using two 70 mm Hasselblad cameras with 80 mm lenses, and a special Air Force 250 mm telephoto lens. He also had a 16 mm Maurer motion picture camera. As they came around the Moon for the fourth time in mid-morning of Tuesday 24 December, Anders was engrossed in taking pictures of the lunar surface, recording the frame information on a tape recorder. Borman kept the spacecraft looking down at the Moon until just before coming into earthview he began to roll the spacecraft for Lovell to take a navigational sighting. Watching the lunar horizon Borman became aware of a colourful patch of blue and white on the edge, rising above the drab gray lunarscape and he realised he was watching Earthrise at the Moon. Startled at the unexpected sight and its breathtaking beauty he exclaimed, "Oh, my God – look at that picture over there; here's the Earth coming up ... wow, is that pretty."

Lovell looked out too, "Oh, man – that's great!"

They all decided the sight had to be recorded on film. Borman grabbed the Hasselblad Anders had just been using and took a picture as Anders jokingly took him to task, "Hey, don't take that, its not scheduled."

Figure 4.42. During Apollo 8, Ken Lee (left) and Station Director Tom Reid man Honeysuckle Creek's Operations Console, the interface between the station and Mission Control at Houston.

As they laughed Borman handed the camera to Anders, but he realised it was loaded with black and white film and there was a scramble to find the colour film.

Anders said, "Let me get it out this window – it's a lot clearer."

Anxious to make sure of the picture Lovell hovered over Ander's shoulder. "Bill, I got it framed – it's very clear right here."

Anders composed the picture in the viewfinder and took two frames as the glowing Earth climbed above the horizon into a jet black sky.

"Oh, that's a beautiful shot," praised Lovell. "Are you sure we got it now?" and suggested Anders take a series of different exposures. "I did – I took two of them." Anders confirmed.

They fell silent and gazed spellbound; the first humans to witness Earthrise at the Moon. The sight of this single orb of rich colour stirred deep emotions. Sent on a mission to study the Moon, the three astronauts found the most profound discovery of the trip was the reassuring existence of their own planet. Borman commented later, "It's a view I will take with me to the grave – that to me was the highlight of the flight."

Anders commented, "How finite the Earth looks. Unlike photographs people see, there's no frame around it. It's hanging there, the only colour in the black vastness of space, like a dustmote in infinity."

Apollo 8 dramatically changed the dimension of the word "home". Originally, the word meant the hearth and the house one was brought up in, then as people moved further afield it expanded to mean the town, then the country one came from. Looking from the Moon back to the Earth suspended in the absolute black, infinite void of space, the astronauts now saw the whole planet Earth as "home". In fact, a running joke among the Apollo 8 crew was that looking at the Earth from space the question arises, "Is it inhabited? Is there life on Earth?"

Just before losing the Earth for the fifth time after Capcom Mike Collins in Houston passed up the Earth's news. Borman settled down to sleep while Lovell and Anders continued working. He slept and rested until the seventh orbit when he talked to his companions for a while.

Lovell asked, "Well, did you guys ever think that one Christmas you'd be orbiting the Moon?"

"Just hope we're not doing it on New Year's Eve," Anders returned.

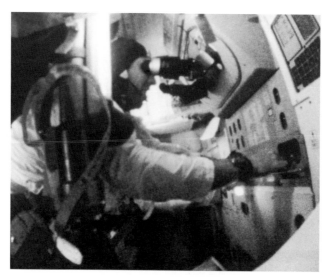

Figure 4.43. Lovell at the navigating station.

"Hey, hey – don't talk like that, Bill. Think positive," the normally chirpy Lovell was serious for a moment.

They studied Apollo landing sites chosen from the Lunar Orbiter photographs, finding some quite unsuitable. Ground controllers picked up a wiggle in the orbit of the spacecraft, which indicated there were gravity changes suspected to come from *mascons*, or "mass concentrations" under the surface of the Moon (for an explanation of *mascons* refer to Appendix 2). These gravity changes affected the Lunar Module coming in to land, so had to be taken into account when planning a descent to the surface.

Houston: "What does the old Moon look like from 60 miles?"

Borman: "The Moon is a different thing to each one of us. I know my own impression is that it is a vast, lonely, forbidding type existence, a great expanse of nothing that looks rather like clouds and clouds of pumice stone."

Lovell: "The vast loneliness up here at the Moon is awe inspiring, and it makes you realise what you have back on Earth. The Earth from here is a grand oasis in the vastness of space."

Anders: "I think the thing that impressed me the most was the lunar sunrises and sunsets. These in particular bring out the stark nature of the terrain. The sky up here is also a rather forbidding, foreboding expanse of blackness."

They saw no colours on the Moon, only a marbling of black, white and various shades of gray. There were endless craters in a land of dirty looking beach sand or what looked like plaster of Paris. It was desolate and forbidding. An estimated half billion people shared a black and white television show of the moonscapes the astronauts were seeing.

NASA officials were rather pleased that some hard, persistent critics of their sending manned spacecraft to the Moon complimented the Apollo 8 crew on their lucid and accurate descriptions. John Dietrich, a NASA geology scientist said, "They have first of all demonstrated their ability to observe from the spacecraft to a degree that I think surprised most of us. It was most encouraging for those of us in the science support area."

By the sixth orbit the crew were so weary Borman called Houston with: "I'm going to scrub all the other experiments, we're getting too tired." Ten minutes later he reported that Lovell was already asleep and snoring.

"Yeah, we can hear him down here," Houston replied.

Houston commented, "There's a beautiful Moon out there tonight," to which Borman replied, "Now, we were just saying that there's a beautiful Earth out there!" This impromptu poem was read up to the Apollo 8 crew from Houston:

> T'was the night before Christmas, and way out in space
> The Apollo 8 crew had just won the Moon race;
> The headsets were hung by the consoles with care,
> In hopes that Chris Kraft soon would be there;
> Frank Borman was nestled all snug in his bed,
> While visions of REFSMATS danced in his head;
> And Anders in his couch, and Jim Lovell in the bay,
> Were racking their brains over a computer display. …
> When out of the DSKY there arose such a clatter,
> Frank sprang from his bed to see what was the matter.
> Away to the sextant he flew like a flash,
> To make sure they weren't going to crash.
> The light on the breast of the moon's jagged crust
> Gave a lustre of green cheese to the gray lunar dust.
> When what to his wondering eyes should appear
> But a Burma shave sign saying: 'Kilroy was here!'
> But Frank was no fool, he knew pretty quick
> That they had been first …. this must be a trick.
> More rapid than rockets his curses they came,
> He turned to his crewmen and called them a name;
> "Now Lovell! Now Anders! Now you don't think I'd fall
> For that old joke you've written up on the wall!"
> They spoke not a word, but were grinning like elves,
> And laughed at their joke in spite of themselves.
> Frank sprang to his couch, to the ship gave a thrust,
> And away they all flew past the gray lunar dust.
> But we heard them exclaim, ere they flew 'round the moon:
> "Merry Christmas to Earth; We'll be back there real soon!"

They celebrated Christmas by reading the Bible, each taking turns at reading the first chapter of Genesis.

Anders began, "We are now approaching the lunar sunrise, and for all the people back on Earth, the crew of Apollo 8 has a message that we would like to send to you: 'In the beginning God created the heavens and the Earth. …'"

Borman concluded the broadcast with: "… and from the crew of Apollo 8, we close with good night, good luck, a Merry Christmas, and God bless all of you – all of you on the good Earth."

Millions of earthlings gathered around their cheerfully-lit Christmas trees with the warmth and comforts of home around them, listened in wonder to this ethereal reading from above that hostile, forbidding lunar surface; so far away in space, so far away from the cosy colours and feelings of home, however humble.

The staff at the Tracking Station at Goldstone were severely hit by influenza. Bill Wood was one member who went down, "On the twenty-third I could hardly drag myself around to make it back to the station. Early on the morning of Christmas Eve my wife Francis said enough was enough and when the car pool showed to pick me up she went out in the dark to tell them I was not going to make it. That day I watched from my bed the crew of Apollo 8 describe what they were seeing. It was only through many long hours by those who were not affected, or those already recovered, that Goldstone was able to support the mission."

Again came an apprehensive moment: waiting to see if the astronauts would have a successful burn to bring them back home. The life and death SPS (Service Propulsion System) motor was designed to be as simple and reliable as possible. It was either on or off. When on, the chemicals ignited on contact and burned full bore, with no regulating controls. The engine had been tested over 5 years and tried 3,200 times without a single failure. A failure would leave them trapped in lunar orbit. As Collins once expressed it: "Swallow the frights one at a time as they appear in the Flight Plan."

Even though it was more than half an hour into Christmas Day, no greetings were passed between the flight controllers in the Mission Control Center at Houston until Apollo 8 was safely on its way home. After 20 hours and 7 minutes in orbit, the rocket engine fired at 1:15 am on Christmas Day, and Apollo 8 aimed for the tiny 48 km wide door into the Earth's atmosphere.

As Apollo 8 emerged from behind the Moon Lovell announced: "Roger. Please be informed there is a Santa Claus."

Marilyn Lovell wore a new mink coat that morning, a Christmas present from her husband. "It came from the Man in the Moon!" she told her friends.

After an uneventful return trip listening to music sent up from Houston and putting on two television shows, Apollo 8 arrived back at the Earth on 27 December, with rapidly increasing speed until it dipped into the atmosphere above Tokyo at a speed of 39,635 kilometres per hour. For the technically minded, the spacecraft guidance system had to aim the Command Module to enter a corridor 48 km wide from the point of the last mid-course correction 310,200 kilometres from Earth at an angle of 6.5° from the local horizontal, within plus or minus 1°.

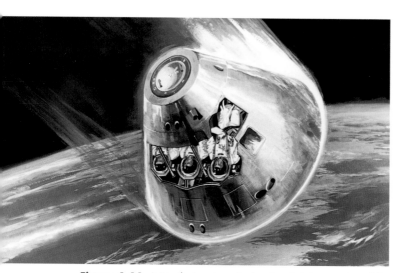

Figure 4.44. A North American Rockwell artist's impression of the Command Module during its fiery return to Earth. Travelling at 39,635 kilometres per hour, it had to enter the atmosphere at an angle of 6.5° from the local horizon, offering a safe corridor only 42 kilometres wide, which allowed an error of only 1° either way.

The resulting fireball, 8 km wide by 161 km long was seen by a Pan American jet airliner flying over the Pacific to Sydney. The Command Module skipped out briefly before entering the atmosphere again for the plunge to the dark Pacific Ocean, far below.

At 7,315 metres the drogue parachutes automatically ejected and slowed the spacecraft down to 225 kilometres per hour for the orange and white parachutes to take over at 3,048 metres to drop Apollo 8 in the Pacific Ocean in the pre-dawn darkness 1,800 kilometres south west of

Hawaii at 9:51 am spacecraft time, or 4:51 am Local time. The recovery carrier *Yorktown's* Sikorsky helicopters only had to fly just 4.8 kilometres to the astronauts and their spacecraft.

The first Apollo mission to the Moon was a triumph for the navigation systems. Launch was 0.6 seconds late, lunar arrival was 0.8 km off target, and after a record breaking voyage of 933,419 kilometres the spacecraft landed a mere 2.6 kilometres from the planned target point.

Saxon said, "We were incredibly lucky at Honeysuckle that out of all the Apollo missions we had the prime role so many times, at so many new, critical and interesting times. In Apollo 8 we were the prime station in view when they first disappeared behind the Moon, and we were the prime station when they appeared from behind the Moon. We were also the prime station for the return to Earth burn. We were also the prime station when they entered the Earth's atmosphere at those incredibly high speeds."

The American Ambassador in Canberra gave the tracking station personnel and their families a special late Christmas party at the Embassy to make up for a disrupted festive season. Over in Spain at the Apollo Tracking Station near Madrid Apollo 8 began an unusual tradition. After splashdown the staff found a suitable restaurant called "Los Bravos" (the Braves) with a suitable atmosphere and old wine cellars in the nearby village of Valdemorillo to celebrate this momentous mission. To their surprise as they stepped through the entrance one of the owner's of the restaurant approached each member of the party with a pair of tailor's scissors and snipped the bottom of their ties and hung them up in a window overlooking the Main Square of the village. The severed ties were never removed and grew in number with each passing Apollo mission.

Staff member José Grandela recalls, "We were attended very respectfully by the people of Valdemorillo who looked at us as wise men closely related with the Universe. They offered us chairs on which they were seated, moved away from the bar to leave us the best places, and retired to a corner just to watch the visitors that spoke a strange language (English), were known by the press and television, and were in daily contact with other men who lived on the Moon."

By rights the names of Borman, Lovell, and Anders should go down in the history books with Magellan, Cook, Amundsen and the Apollo 11 crew, as they were the first to leave the gravity of Earth and arrive at another celestial body, the Moon. The Apollo 8 Command Module is on display in the Chicago Museum of Science and Industry, Chicago, Illinois.

1969 The Crew to Land and Walk on the Moon is Chosen

Following the successful Apollo 8 flight rumours began to fly around that these three would be chosen for the first lunar landing, but on 6 January 1969 Deke Slayton called the back up crew for Apollo 8, Neil Armstrong, Buzz Aldrin, and Michael Collins into his office and said to them, "You're it, guys. You've got the Apollo 11 flight, and that means you get first crack at landing on the Moon. That is, of course, if we pull off successful missions with 9 and 10." On 8 January 1969 the names of the crew of Apollo 11 were announced to the public.

Following on from the Gemini procedures, where the pilot climbed out to walk in space, it had originally been planned for Buzz Aldrin to be first to step out on the Moon. However when it was actually tried out in a mock up they found that Aldrin had to climb over Armstrong to get to the hatch and their bulky spacesuits and back packs damaged the LM, so it was decided that Armstrong would climb out first. Also Armstrong had joined the space programme before Aldrin so had seniority, to which Slayton added, "Secondly, just on a pure protocol basis I figured the Commander ought to be the first guy out."

Did Armstrong pull any strings?

"Absolutely not," said Slayton.

"I was never asked my opinion," said Armstrong.

"It was fine with me if it was to be Neil," said Aldrin at the time.[17]

1969 The Russians Succeed in the First Crew Transfer in Space ...

The Russians were still active. At 10:30 am Moscow time on 14 January 1969 Soyuz 4 took off from Baikonur with Vladimir Shatalov on board. The next day at 10:05 am Soyuz 5 launched Boris Volynov, Aleksei Yeliseyev, and Yevgeni Khrunov into space to rendezvous with Soyuz 4 for about three orbits or $4\frac{1}{2}$ hours. At an apogee of 251 km and a perigee of 211 km, Yeliseyev and Khrunov transferred from Soyuz 5 to join Shatalov in Soyuz 4, and the Russian programme had chalked up another space first, the transfer of crews between two spacecraft. Soyuz 4 landed on the 17 January at 9:51, and Soyuz 5 at 10:58 am on 18 January.

1969 ... But Lose their Biggest Moon Rocket

With their satellites the Americans were carefully watching the Soviet launching facility at Baikonur and had spotted a huge rocket being prepared for flight, this time an unmanned trial flight. The Americans sat up and took a closer look – it was bigger than the Saturn V so it had to be destined for the Moon. It was the Series 3L, N-1 rocket, Korolev's answer to the Saturn V. In May 1960 Korolev had proposed to Premier Khrushchev that Russia could have a big rocket quite soon and submitted a proposal which included an input by rival designer Vladimir Chelomei. Although there was some opposition by the other rocket designers, the plan was approved by Government decree on 23 June 1960, and work began in earnest. Korolev died before the N-1 was ready and Vasily Mishin, his successor, personally led 2,300 technicians in round the clock shifts to prepare the N-1 for this, its first flight. It looked like the race for the Moon's surface was getting serious. The Americans had sent three men around the Moon but still had two missions to go before actually landing. Would the Russians try to beat the Americans by landing at their first attempt if this monster rocket on their launch pad flew successfully?

At 12:18 pm Moscow time on 21 February 1969 the thirty engines of the N-1 fired and lifted the largest rocket ever built, tall as a thirty seven story building, off the launch pad. As it accelerated into the atmosphere two engines shut down prematurely while the rest went through their computer controlled routine. At sixty seconds into the flight when full throttle was applied again after going through the shock barrier, the vibration shook a fuel line off and spewed super cold liquid oxygen into the bowels of the rocket. Detecting a fire the engine monitoring systems shut the motors down and range safety destroyed the rocket at 70 seconds from lift-off.

The resulting fire and explosions were awesome to watch as the blazing colossus spread into a fiery red mushrooming cloud in the night sky and showered back to Earth as burning, smoking shrapnel, 45 kilometres away.

Any Russian hopes of landing on the Moon before the Americans faded away with the dying embers of the fireball they had just witnessed. Although another N-1 launch was tried before Apollo 11 this disaster effectively ended the struggle between the two super powers for being first to land on the Moon. There was no way the Russians could recover from this disaster before America's astronauts stepped onto the Moon's surface – if NASA's plans went according to schedule.

1969 Apollo 9: Testing the Apollo spacecraft

Apollo 9 was the first time that the whole Saturn vehicle was to fly complete, so all the lunar manoeuvres could be rehearsed with the safety of Earth at hand. The crew of Apollo 9 worked so hard for so long to meet their launch date, they all got upper respiratory infections, and the launch had to be delayed three days to a lift off from Pad 39A at 11:00 am EST on 3 March 1969.

Once in orbit the Command Module had to turn around to pull the Lunar Module out of its Saturn IVB "garage". To dock the two spacecraft together the astronauts had to use a complex system of latches, probes and drogues called the docking adapter. The probe was a tube extending out of the nose of the Command Module, and the drogue a hole in a hatch of the Lunar Module. Once the probe was inserted in the drogue it would retract and pull the two spacecraft together and a series of twelve latches would lock and allow the probe to be removed to let the astronauts float between the two spacecraft. If the system did not work it meant the end of the lunar landing, and if it did not work coming back after a landing, the returning crew would probably not make it back alive

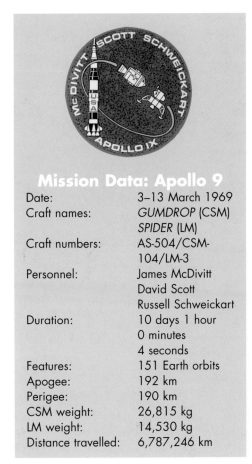

Mission Data: Apollo 9	
Date:	3–13 March 1969
Craft names:	GUMDROP (CSM) SPIDER (LM)
Craft numbers:	AS-504/CSM-104/LM-3
Personnel:	James McDivitt David Scott Russell Schweickart
Duration:	10 days 1 hour 0 minutes 4 seconds
Features:	151 Earth orbits
Apogee:	192 km
Perigee:	190 km
CSM weight:	26,815 kg
LM weight:	14,530 kg
Distance travelled:	6,787,246 km

The first two days were spent testing the docking equipment, dropping off the SIVB rocket and running trial of the SPS motor. All were completely successful, so on the third day in space Rusty Schwieckart first entered the Lunar Module at 7:27 am to begin preparing for its first free flight and trials. Schweickart's friends called him "Rusty" because of the hue of his hair, but on this mission he was code named *Red Rover* for his spacewalk. The Gemini spacesuits had relied on the spacecraft to supply all the oxygen and communications; the new lunar spacesuits were completely independent of the spacecraft but still had to be tested in space. The first problem of the mission was Schwieckart getting sick after climbing into the new pressure suit. His condition caused some concern on the ground so they spent the whole day quietly in the Lunar Module and just opened the hatch for Schwieckart to climb out onto the porch platform while they were above the Pacific Ocean. He slipped his boots into gold painted restraints, looked down and exclaimed, "Boy, oh boy – what a view!"

The fifth day was the big action session when *Spider* and *Gumdrop* separated by 161 km and over six hours and twenty minutes rehearsed the landing and launch sequences,

Figure 4.45. How the astronauts first see the Lunar Module in space. Here *Spider* is nestling in the Saturn V third stage with its legs folded, in position for the launch sequence. The protecting covering panels have been blown off so the Command and Service Module can now dock and pull the Lunar Module out and away from the Saturn IVB rocket.

jettisoned the Lunar Module's descent stage, and practiced rendezvous and docking procedures to be performed by the upcoming lunar landing missions. It was the first time that astronauts had flown a spacecraft that could not return to Earth as the Lunar Module had no heat shield – they *had* to return to *Gumdrop* to get back home.

Scott welcomed them back with, "You're the biggest, friendliest, funniest looking *Spider* I've ever seen."

When the buzzer sounded to indicate the two spacecraft were docked safely, McDivitt sighed with relief, "Wow! I haven't heard a sound that good for a long time."

Don Gray, then Station Director at Tidbinbilla, explained, "The Apollo 9 mission had been planned for the Australian end that Honeysuckle Creek would track the Command Module and Tidbinbilla the Lunar Module. There was a problem from Tidbinbilla's point of view, its antenna was much slower than Honeysuckle's as it was a deep space station, most of its life only needing to move at sidereal rates. But there came a period where there was a requirement for both the Command Module and Lunar Module to be tracked while they were separated, and in particular they needed to get commands into the Lunar Module. Honeysuckle Creek's antenna couldn't see both spacecraft at the same time and the Tidbinbilla antenna wasn't capable of following either of them at the high speeds in Earth orbit.

"We sat down at Tidbinbilla – one of the things we had on our antenna was an acquisition antenna with a very much wider beamwidth than the main dish, we also had a pretty good high powered transmitter and so we came up with a scheme and put it to NASA that we would use that acquisition antenna with its very wide beamwidth – we would drive the antenna under computer control from horizon to horizon at its maximum possible rates across the sky – we would catch the Lunar Module in the leading edge of the antenna beam – the spacecraft would fly through the antenna beamwidth and out the other side. The Lunar Module would be in the antenna beam long enough for the required commands to be blasted in. NASA accepted that as the operational procedure and that, in fact, was one of the best things we ever did – we *knew* our equipment well enough in Australia to be able to make those proposals to NASA, and they trusted us well enough to accept our word for it – mind you, they did send an aircraft out here and check it out for hours on end to see that we could do it."

Apollo 9 splashed down in the Atlantic at 11:00 am Houston time on 13 March to be recovered from the ocean by the USS *Guadalcanal*. The mission was declared a success, so the way was now clear for the final assault on the Moon, providing the Apollo 10 dress rehearsal did not spring any surprises.

Figure 4.46. Dave Scott standing in the Command Module *Gumdrop*'s hatch during the fourth day of the mission. The picture was taken by Rusty Schweickart from the porch of the docked Lunar Module *Spider*.

Figure 4.47. This cyclonic storm system, 1,900 kilometres north of Hawaii, provided this spectacular sight as Apollo 9 cruised past on their 124th revolution. For the first time in history, spaceflights have given mankind the opportunity to see whole storm systems, cyclones and hurricanes at a single glance from above.

Figure 4.48. Russell Schweickart, Dave Scott and James McDivitt acknowledge the USS *Guadalcanal*'s greetings.

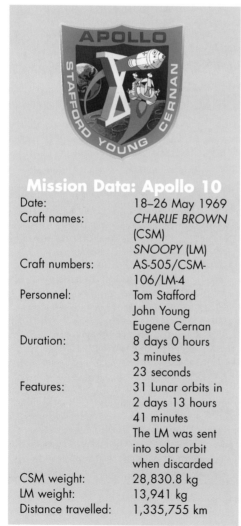

Mission Data: Apollo 10

Date:	18–26 May 1969
Craft names:	CHARLIE BROWN (CSM)
	SNOOPY (LM)
Craft numbers:	AS-505/CSM-106/LM-4
Personnel:	Tom Stafford
	John Young
	Eugene Cernan
Duration:	8 days 0 hours 3 minutes 23 seconds
Features:	31 Lunar orbits in 2 days 13 hours 41 minutes
	The LM was sent into solar orbit when discarded
CSM weight:	28,830.8 kg
LM weight:	13,941 kg
Distance travelled:	1,335,755 km

1969 The Final Dress Rehearsal: Apollo 10

Cernan explained, "For names we came up with something people could relate to, particularly kids. Everybody knew *Snoopy* and *Charlie Brown*, they were characters that both young and old knew. Maybe we were a product of the times – they were very popular characters, not very sophisticated names as you saw later in Apollo 17. I don't remember who came up with it, maybe it was John Young, maybe myself – maybe both of us. *Snoopy* was appropriate because it was going to do a lot of snooping around the surface of the Moon. It humanised the space program – the technology was so far above everybody. We got lots of letters on *Snoopy* and *Charlie Brown*."

Under normal circumstances the crew of Apollo 10 would appear to be the team to land, but there were a number of reasons to make it the penultimate mission. Not enough was known about the Moon's environment for some of the manoeuvres, for example there were gravity peaks caused by heavy material under the Moon's surface called *mascons*, which affected a spacecraft's flight path; the lunar landing computer software wasn't quite ready; and the Lunar Module was a shade overweight which may have caused problems lifting off the lunar surface. So it was planned that the Apollo 10 Lunar Module would fly within 15,240 metres of the lunar surface, and sink into oblivion instead of celebrated world history.

At 12:45 pm EST on 18 May 1969 Apollo 10 left the Kennedy Space Center and headed for the Moon on the second Earth orbit. This was the only time that Pad 39B and firing room 3 were used for an Apollo mission. During the trans-lunar injection burn the astronauts were sud-

Figure 4.49. The human touch invades a bastion of technology. Popular characters Charlie Brown and Snoopy from the comic strip "Peanuts" ride the consoles at the Mission Control Center during Apollo 10.

denly shuddering to vibrations from the booster rocket pressure relief valves, and as their vision began to blur they all feared the mission was going to end before they had left the Earth. Stafford's fingers reluctantly curled around the abort handle as he called Houston through gritted teeth, "Okay, we are getting a little bit of high frequency vibrations in the cabin." Luckily, after five minutes of suspense the burn ended on time and they were safely on their way to the Moon, later trying out the first colour television camera to be used on a lunar flight. They went behind the Moon at 4:45 pm on 21 May, to emerge on the other side full of the exciting views of the Moon they were seeing.

Apollo 10: "Houston, Ten."

Houston: "Go ahead, Ten."

Apollo 10: "It's amazing what you can see with earthshine on the surface of the Moon – it seems to be very well lit from our altitude here. The Moon past the terminator is totally dark as long as we are in sunlight, but the minute we go out of the sunlight into darkness ourselves, the Moon then glows right at us."

Cernan: "The LM thrusters stick out like a sore thumb in earthshine, but they don't keep you from seeing any of the stars at night – it's real well lit up."

Stafford: "In earthshine you can see right into the craters, you can see shadows in the craters from the earthshine. The more you become adapted to it, it's phenomenal the amount of detail you can see."

Houston: "Roger Ten."

Cernan later recounted, "When we opened the tunnel hatch a lot of the LM's insulation had come loose – it was like a snow storm in there – we had a lot of problems with that for a while."

Stafford added, "There was a big thermal cover on the front of the Command Module filled with fibreglass, and when we opened up the tunnel to pressurise the Lunar Module the aluminium cover ripped and all the fibreglass just blew out as the air flowed in and the LM just filled full of the stuff. Gene had the stuff in his eyebrows – his hair looked like he'd just come out of a chicken coop. Then we all started to itch – we itched all the way back. It eventually went to the inlet screens of the air-conditioning and we tried scraping it off there. We still had pieces of it all the way back."

Following some concern about an unusual twist of 3.5° in the two spacecraft's alignment while docked due to some holes that hadn't been drilled, *Charlie Brown* and *Snoopy* separated. Stafford and Cernan were now in a spacecraft that could not get them back onto the Earth. They were relying on the Lunar Module giving a faultless performance and their being able to rendezvous and dock safely. What if something went wrong on their return to Young, and they were unable to dock properly, as nearly happened on

Figure 4.50. Tom Stafford's view from *Snoopy* of a gleaming *Charlie Brown* coming in to dock on the far side of the Moon. The pristine shine of the Command Module was soon obliterated during the fiery re-entry into the Earth's atmosphere.

Apollo 14? They would be stranded in lunar orbit, and Young would have to return home alone.

As *Charlie Brown* fell away, Young called, "Adios, we'll see you back in about six hours."

Cernan replied, "Have a good time while we're gone, baby."

Stafford added, "Yeah – don't get lonesome out there, John."

They fired *Snoopy*'s rocket to drop down to within 15,240 metres of the Sea of Tranquillity. Looking down from high above Young reported, "They are ramblin' among the boulders."

As they made their first pass over the southwestern corner of the Sea of Tranquillity an excited Cernan called out, "I'm telling you, we are low. We're close baby! We is down among 'em, Charlie."

Capcom Charlie Duke responded, "I hear you weaving your way up the freeway."

Snoopy raced across the face of the planned prime Apollo 11 landing site at just under 6,000 kilometres per hour, Stafford giving a running commentary on the features he could see out the window and trying to take pictures with a faulty Hasselblad camera. They checked the computers, landing radar, and kept an eye open for any unexpected surprises.

They then looped back out to return for the second pass when they planned to simulate a launch above the landing point by firing the ascent stage rocket to separate from the descent stage and return to *Charlie Brown*.

As Stafford brought the Lunar Module down for the second pass he spoke to Houston, "Okay, we are coming up over the site. There's plenty of holes there. The surface is actually very smooth, like a very wet clay … with the exception of the very big craters."

At 7:32 pm they were preparing to stage, to drop the descent part of the Lunar Module off, when without any warning the two astronauts found themselves spinning around out of control as *Snoopy* began to jerk and buck about like a wild horse.

Stafford recalled, "We were going along upside down and backwards about a minute before the descent stage thrusters were due to fire, I looked down and the instruments were showing we were yawing right, and I looked at the eight ball and I had no yaw, so it looked like we had a yaw right gyro failure and it also looked like an electrical glitch, so I started trouble shooting, but we had a wrong position on a switch and the whole thing started tumbling over at about 6° per second. I just reached over and blew off the descent stage early. All the attitude control thrusters were on the ascent stage and the descent stage weighed about twice as much as the ascent stage so without the descent stage I had more control."

Cernan remembered, "I saw the lunar horizon coming through my window about five times from different directions in about 8 seconds. We were able to throw the right switches to get it under control – it was hard but we did it. We weren't being banged around the cockpit, we were cinched down in our straps in a standing position."

Cernan had said during the mission, "Son of a bitch! I don't know what the hell that was, baby. The thing just took off on us. I tell you, there was a moment there. … I thought we were wobbling all over the sky."

He later said, "Those words came involuntarily – I didn't even know I had said them until I got home and somebody played the tape for me. It was just a natural emotion – those words didn't come out until after we got things under control. You know, you're a long way from home at 50,000 feet above the lunar surface and all of a sudden you're spinning around in three different directions – 8 seconds can be a long time."

Acutely aware of the menacing terrain racing past their windows, with mountains grinning at them like gigantic decayed teeth as Cernan saw it, Stafford took over manual control and *Snoopy* promptly quietened down. A few more seconds out of control and they would have tumbled onto the lunar surface to be splattered over the craters. The wild gyrations were caused by a switch in the navigation system set incorrectly. The Lunar Module had two computers; the Primary Navigation Guidance System (PNGS, pronounced "Pings") for indicating where they were, and the Abort Guidance System (AGS) for use near the Moon in an emergency launch. During the checks Cernan had correctly switched from PNGS to AGS, then unaware the switch had already been operated Stafford had switched it back to PNGS so instead of keeping *Snoopy* on a steady course, the system lost control looking for *Charlie Brown* that wasn't there. To add to the confusion Stafford then threw the switch back to AGS thinking it was going to PNGS.

"I was on the right hand side," said Cernan, "and we were using both the Primary Guidance computer and the Abort Guidance computer for test purposes for that staging event and I put the switch in a specific position for Tom to go ahead and stage and Tom didn't know I had put it there, so he moved it back to the other position. It all happened so fast we weren't sure how it happened. I was always convinced that it was a switch that had been mistakenly, but legitimately, moved by each of us, both trying to do the right thing."

Ten minutes later the rocket fired to push them back up into orbit to meet Young in *Charlie Brown*. At 11:00 pm they met, and Mission Control broke out a large cartoon showing *Snoopy* kissing *Charlie Brown*, saying, "You're right on target, *Charlie Brown*," and a little later *Snoopy*'s ascent engine was fired for a test before it was cast off.

Cernan said, "We sent *Snoopy* off into orbit around the Sun – I don't know if it's still there or not. Who knows? Who knows where that thing is now?"

The Fastest Humans in History

After 31 orbits Apollo 10 and its crew left the moon, sending TV pictures of it back to Earth. With extra fuel left over from the lunar activities they burnt it off to accelerate the spacecraft back to Earth at a record 39,897 kilometres per hour, making the crew of Apollo 10 the fastest humans in history according to the Guinness Book of Records[57]. However, according to data supplied to the author by Rod Rose, past steward of the NASA records, Apollo 10's speed during Entry Interface at 191 hours 48 minutes 54 seconds

Figure 4.51. The Australian Commonwealth Scientific and Industrial Research Organisation's (CSIRO) 64-metre radio telescope at Parkes, New South Wales, was commandeered to support the Apollo 11 mission to maximise the quality of the signal from the lunar surface. It produced the best quality pictures of the Apollo 11 moonwalk.

Ground Elapsed Time was 36,315 feet per second, or 24,760.227 statute miles per hour (39,846.60 kilometres per hour).

At 11:52 pm Houston time on 26 May, they splashed down into the Pacific 635 kilometres east of Pago Pago, 5 kilometres from the recovery carrier USS *Princeton*. The crew aboard the carrier were treated to a spectacular sight of the Service Module streaking across the pre-dawn sky in a blazing fireball as it burned up, followed by the Command Module silhouetted against the brightening sky under its three big parachutes. Waiting beneath, the recovery helicopters buzzed about with their flashing running lights stabbing the dark velvet blue sky. When the crew in the spacecraft looked up at the helicopter hovering above they saw "Hello there Charley Brown" written across the bottom.

Cernan explained, "Was it frustrating to be so close to landing? – absolutely not. Just to be on a flight that went that far to the Moon – there was so much at stake on our flight – we almost dictated whether Apollo 11 was going to get the chance or not. Even twenty five years later in retrospect it was not frustrating, nor was it disappointing to me. Of course, I had the chance to go back, which was important to me as well."

The CSIRO's Radio Telescope at Parkes in New South Wales, Australia, Joins the Apollo Tracking Network

During the years following the Second World War, Australia's CSIRO Radiophysics Laboratory led the world in radio astronomy, and to capitalise on this success built a 64 metre dish antenna at Parkes, 210 km west of Sydney, for serious radio astronomy research. The celebrated British inventor Barnes Wallis was involved in the early design of the Parkes telescope, and some of its design features were used as a basis for the design of the original JPL big Deep Space antennas, one of them at Tidbinbilla.

Officially opened by the Governor General, Lord De L'Isle, on Tuesday, 31 October 1961 the new radio telescope quickly leapt to the forefront of cosmic exploration,

Figure 4.52. Personalities at Parkes. CSIRO scientist Dr Paul Wild (left) and the then Director of the Parkes Radio Telescope, Dr John Bolton (right) with Roy Stewart from the NASA Deep Space Station at Tidbinbilla.

being involved in the discovery of quasars. The Parkes radio telescope was about to add another illustrious phase to its career – supporting the Apollo 11 mission.

Dr John Bolton, Director of the Parkes Observatory at the time, explains how it happened, "I was approached in 1968 in the USA at a carefully-arranged dinner party at Bob Leighton's home at which the only other guests were Eb Rechtin of the Jet Propulsion Laboratory and his wife. We were asked to back up Goldstone for the planned Moon walk, in case of a delayed take off or other reasons. The fact there were to be human beings at risk in space sold the project to Taffy (Dr E.G. Bowen, originator of the Parkes Telescope) and me but, even though there was to be some reimbursement for the Radiophysics coffers, it found little favour with the staff. Rechtin made a visit to Parkes shortly afterwards.

The next step was a meeting convened with the Australian Department of Supply to arrange contract details. For two hours this meeting addressed neither Bob Taylor, the American engineer to be Parkes based, or me – the principal actors. Fortunately he and I had spent the previous evening together discussing our roles, so I ended the meeting by saying that Taylor and I could work together and that I would only accept a one line contract – the Radiophysics division would agree to support the Apollo 11 mission."[41]

Special Links Installed at Honeysuckle Creek for Live Coverage of the Lunar Landing

At this point there was no provision to transmit a television picture from Honeysuckle to the outside world, and it was decided to install a television link from Honeysuckle Creek to Canberra. Trevor Gray was a Post Master General's Department technician on the installation team. "We worked hard night and day to get this link in place. First a temporary tower was put up beside the road into Honeysuckle, and dish antennas mounted on top. Equipment was then connected to them, most of it in the basement of the station building. The link went from Honeysuckle to Williamsdale, then back to Red Hill in Canberra.

"We had mixed up brand dish antennas – everything was grabbed from everywhere. We were told there was no money for capital works – we couldn't buy anything, but there was plenty of money for operations. There was some Collins gear that NASA had, and we had to match their gear to our gear – so someone showed us a lathe out the back of Honeysuckle and having had a few hours of experience on lathes before we made these joints up and bolted them together. For the actual Moon landing the ABC (Australian Broadcasting Commission) put in another link in beside ours, so we had two links plus a standby.

"These links were difficult to maintain – it was wintertime and very cold. There was a high voltage in these joints and moisture got in them and a few of them blew up, naturally some in the middle of the night. We ended up sealing them up with epoxy resin. Normally these links were designed to only be up for a hour or so, say during a football match, they weren't meant to stay there for long. By the time of the first Moon landing we had it settled down enough to last through the mission. It meant that not many of us got much sleep."

The Russians Experience Another N-1 Booster Rocket Setback

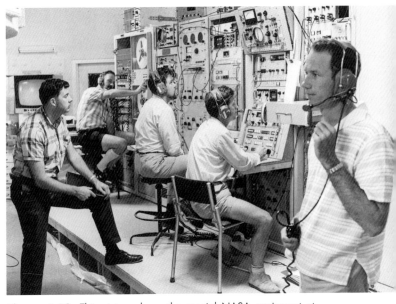

Figure 4.53. This picture shows the special NASA equipment at Parkes manned by staff from the Deep Space Station at Tidbinbilla under Bruce Window (right). This equipment was kept until the end of the Apollo Program.

The world's media was now looking to the United States and gearing up for that historic, epoch making event – Apollo 11 and a human stepping on-to another celestial body. Despite their recent N-1 catastrophe the Russians were still busy. Just two weeks before Apollo 11 left the Earth, the Russians tried another N-1 launch, planned to send a space-craft on a loop around the Moon. The launch teams were still working round the clock using hastily-prepared sketches and red-lined drawings to prepare the next N-1 (Series 5L) for launch. On 3 July 1969 the thirty motors fired and almost straight away the oxidiser pump of motor #8 exploded from digesting a loose metal fragment. A fire was already burning inside the rocket before it had cleared the tower and the rest of the motors were shut down. Then, as if in slow motion, the rocket leant to an angle of 45° and dumped itself back on the pad. The resulting liquid fireball and explosion completely destroyed the launch complex, taking another practice N-1 stored nearby with it. The shock waves were reported to have been recorded 3,219 km away in Stockholm, and sur-veillance satellites showed a big hole in the ground where launch complex 110R had been.

It was two years before the Russians tried another N-1 launch, which also failed.[35]

As a final gesture of defiance, the Russians sent their Luna 15 spacecraft up on a Proton booster just three days before Apollo 11 left the Earth. It entered lunar orbit during 17 July, and stabilised into a 111 × 16 kilometre orbit, taking 1 hour 54 minutes to go around once. The Russians had planned it to land on the surface, drill a core sample, and bring it back to Earth before the Apollo 11 astronauts returned, to try and convince the West they were not racing to put a man on the Moon; and that a robot can do the job just as effectively as man. Luna 15 crashed onto the lunar surface, and it was left to Luna 16 to bring a sample back – but by then it was too late – Apollo 11 had already brought back over 20 kilograms of lunar rocks.

1969 **NASA: "Now we know we can Land on the Moon"**

The Americans were now ready for a landing on the moon. NASA Administrator Paine said, "Today, this moment, with the Apollo 10 crew safely on board, we know we can go to the Moon."

The Americans had been right down the track, right down to the last step. All the paving blocks to complete the track to the actual moon's surface were set firmly in place.

Armstrong, Aldrin and Collins, watched Apollo 10 very closely, and were very relieved when it was declared a success as it meant that their Apollo 11 mission would be the first attempt to land. Compared to Alan Shepard's 150 hours of simulations for his first Mercury flight, they each spent over 400 hours, fifty working days, in simulators wrestling with a continuous stream of missions, usually peppered with emergencies, equipment malfunctions, and potential catastrophes to test their knowledge, skill, and coolness to the limits. In the simulator, or the Great Train Wreck as John Young preferred to call it, the training supervisors could fail any subsystem and any of the bank upon bank of 678 switches and 410 circuit breakers – including toggle switches, thumbwheel switches, press button switches, rotary switches, switches with locks, switches with interlocks, circuit breakers, rods, levers, and control sticks sending signals through 64 kilometres of wire.

The astronauts constantly checked dials, meters, lights, alarms, and indicators looking for signs of trouble that could be introduced by the simulation engineers.

They had to understand, operate, check, control, and monitor parts of the spacecraft such as environment control, fuel, propulsion, photography, radiators, radars, sextants, antennas, timers, computers, purge valves, navigation instruments, pressure regulators, temperature controls, medical monitors, communication panels, fan motors, food, sanitation, lights, fuel cells, guidance systems, telescopes, and waste water systems.

The astronauts entered the computer-driven sim-

Figure 4.54. Neil Armstrong steps down from the Flight Simulator after a training session. The Apollo 11 crew spent over 400 hours training for the mission.

ulators prepared for anything, any situation the training supervisors could dream up to exercise their victims to the limit. Hours later they usually crawled out of the cockpit stressed out in a ball of sweat from a continuous bombardment of simulated life threatening situations and equipment failures designed to find any weaknesses in their operational and technical skills. But – by the time they were ready for going to the moon, the astronauts knew every twist and turn of the normal and emergency operational procedures, every capricious component of the spacecraft's 26 subsystems.

Commander of the Apollo 11 mission, Neil Armstrong, "We had a great deal of confidence. We had confidence in our hardware: the Saturn rocket, the Command and Lunar Modules. All flight segments had been flown on the earlier Apollo flights with the exception of the descent and ascent from the moon's surface, and of course, the exploration work on the surface. Although confident, we were certainly not over confident. We were not overly concerned with our safety, but we would not be surprised if a malfunction or an unforeseen occurrence prevented a successful landing. As we ascended the elevator to the top of the Saturn we knew hundreds of thousands of Americans had given their best effort for us.

Now it was time for us to give our best."[42]

Dr Robert Goddard: "It is difficult to say what is impossible, for the dream of yesterday is the hope of today and the reality of tomorrow."

5 Apollo 11

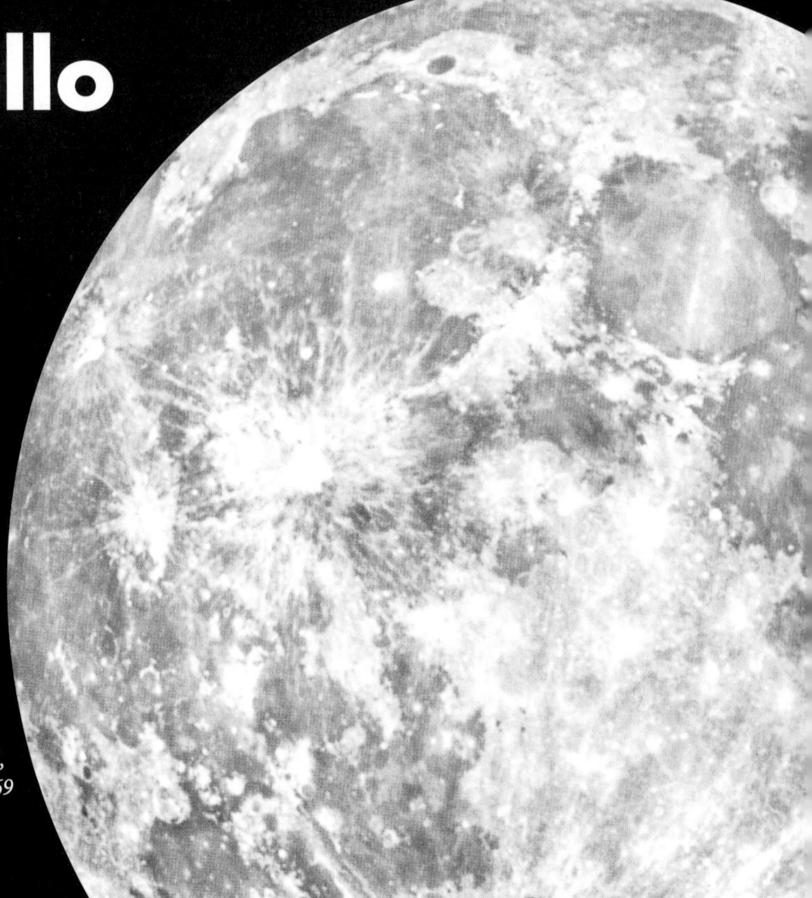

The lunar landing of the astronauts is more than a step in history; it is a step in evolution.

New York Times Editorial,
20 July 1969

The Apollo 11 Astronauts

Director of Flight Crew Operations, astronaut Deke Slayton, had the job of selecting the crew for Apollo 11. With Apollo 9 and 10 missions still to come and no guarantee that Apollo 11 would do the first landing there were a number of candidates for the crew, topped by Frank Borman and Jim McDivitt, followed by Tom Stafford, Neil Armstrong, and Pete Conrad. Originally Slayton felt that Gus Grissom would have been his first choice, but with his death it was tempting to put Borman on with his experience from Apollo 8. But Borman had already indicated he was ready to pull out because of the stress on his family. There was also talk of sending Apollo 10 down to the surface, but that was ignored, and Slayton decided to stick to the normal routine of rotating the back up crews, which put Neil Armstrong, Buzz Aldrin, and Fred Haise on the Apollo 11 mission; but there was a change. Michael Collins was back on flight status and looking for a berth, which reminded Slayton he had originally promised him a lunar landing flight. As Haise hadn't figured in the original plans, Collins joined the crew as the Command Module

◀ **Figure 5.1.** The dream comes true. Buzz Aldrin standing beside the Lunar Module among the first human boot prints in the Moon. For the rest of time we can think of ourselves as from the Planet Earth.

pilot. This fitted in quite well as Aldrin had already trained on the Lunar Module before the Apollo 8 to 9 crew swap so he moved across to become the Lunar Module pilot, and the Apollo 11 crew became Neil Armstrong, Commander, Michael Collins, Command Module pilot, and Buzz Aldrin, the Lunar Module pilot. The back up crew would be Jim Lovell, Bill Anders, and Fred Haise.[29]

Launch Preparation

Dr Kurt Debus, Director of the Kennedy Space Center and once a war-time V-2 launch director in Germany, finally announced, "When the weight of the paperwork equals the weight of the Stack – it's time to launch!"

Actually, the time to launch Apollo 11 was carefully chosen so that the Lunar Module would land on the Sea of Tranquillity with the morning Sun low enough to throw strong shadows on the lunar surface for the best feature manifestation from above. The mission planners set 16° as the highest Sun angle for a landing and 6° the lowest, and judged that anything greater than 20° was unsatisfactory for a manual landing. Also, Apollo 11, as with the other lunar missions, was sent to orbit around the Moon in a clockwise direction so that the Sun would be behind the Lunar Module, and not in the astronauts' eyes as they came in to land.

Mission Data: Apollo 11	
Date:	16–24 July 1969
Craft names:	COLUMBIA (CSM)
	EAGLE (LM)
Craft numbers:	AS-506/CSM-107/LM-5
Personnel:	Neil Armstrong
	Michael Collins
	Edwin "Buzz" Aldrin
Duration:	8 days 3 hours 18 minutes 35 seconds
Features:	First manned landing on the Moon
EVA:	2 hours 31 minutes 40 seconds
Lunar samples:	21.55 kg
CSM weight:	28,807 kg
LM weight:	15,095 kg
Landing area:	0.688° N, 23.433° E on the Sea of Tranquillity
LM impact:	Unknown
Distance travelled:	1,534,832 km

A lunar day lasts two Earth weeks, so it takes seven days for a morning to reach a lunar midday when the temperature soars to 110°C. During the two week lunar night the temperature plummets to minus 170°C.

Armstrong explains the choice of landing time: "The primary reason was that we wanted to land early in the morning – that is when the Sun was at a fairly low angle, between 5 and 15 degrees above the horizon, so the temperatures wouldn't get to us and also so the lighting would be good in that we would have quite a bit of shadows in the landing area so our depth perception would be as good as possible.

"The Moon has a very peculiar lighting characteristic. If you walk out in your backyard and look at the Moon, it doesn't have a bright spot on it. If you look at a Christmas tree ball, it has a shiny spot where you have a direct reflection of a light source and this is called specular lighting. But if you look at the Moon from your backyard it is almost equally illuminated. This characteristic is further represented by the fact that light comes right directly back into your face from the Sun. All of you who have flown close to clouds or low to the ground over dewy fields know that when you look at your shadow it has a bright halo around it. That particular characteristic is very much larger on the Moon. You have a very bright spot right along the Sun line shining back into your eyes. It's so bright that we were considerably worried about it and we wanted to choose the final descent so that we wouldn't be looking into this bright spot of light right down our flight path. So, these various considerations boxed us into a fairly tight approach angle and we wanted to be fairly close to where the shadow was so that the Sun would be only about 10 degrees above the horizon."[44]

Figure 5.2. The Ship: the Command and Service Module *Columbia* ready for mounting on top of the Saturn V in the VAB.

The night before the launch, lights burned late in Cape Canaveral, in Houston, and around the world the operational, technical, recovery and rescue teams prepared their equipment and the vast, complex, communication networks supporting the flight.

The Network Controller from the Goddard Space Flight Center briefed the world wide tracking network of 30 stations, 4 ships and 8 aircraft just before the mission: "The chances of hardware problems in the spacecraft and on the ground which could seriously jeopardise the mission's success were much less than the chances of a person pushing the wrong button at the wrong time. For example, unless the antenna is pointed at the right place at the right time the station might as well not be there.

"Also, when the antenna technician has done his part, the transmitter and receiver technician must push the right buttons at the right times if any data or voice up or down to the spacecraft is to be received. This operational performance requirement follows down the line, i.e. the chain is as strong as its weakest link.

My role in this mission, as Operations Supervisor, is to co-ordinate station planning and operations to make each link of the chain as strong as possible, and to efficiently use parallel chains wherever available, so that a possible failure in a link would be covered. Being as involved in the detail of the station it is very easy to lose sight of the overall picture – somewhat analogous to a stage manager who has to be aware of all details of the production, but never sees the end product as does the audience.

"I see this mission as being only a further step in man's inevitable curiosity and drive to explore, and look forward to taking a part in future steps of this kind."

With an event such as the Apollo missions, a lot of our daily preoccupations are no longer applicable to travellers in space. Once they have left the Earth the weather, for instance, is of no interest to the astronauts as they have no weather, and they can select any of Earth's time zones for their day as they have no sunset or sunrise. Unless otherwise noted, the times quoted in this Apollo 11 story are Houston Daylight Saving Time used by the astronauts on the spacecraft, so the reader can relate the activities to their own normal day on Earth. The mission was planned around a Ground Elapsed Time, or GET as it was known in the industry, which began counting up in hours from the moment of launch. The moment the vehicle left the pad, the GET and all the planned events could be related to any time zone on Earth.

Getting Ready

Launch day, Wednesday 16 July 1969, dawned with a few fleecy clouds, and the promise of a hot, sunny day. The terminal countdown had already begun at T-minus-28-hours at 5 pm Florida time on 14 July. With two built-in holds of 11 hours and 1 hour 32 minutes, it gave a planned launch time of 9:32 am EDT on the 16th. The fuelling of the launch vehicle was completed more than three hours before liftoff when the closeout crew of six men under the direction of Gunter Wendt and Spacecraft Test Conductor Clarence Chauvin returned to the vehicle and began the final cabin preparations.

The three astronauts began the day at 4:15 am EDT with a breakfast of orange juice, steak, scrambled eggs, toast, and coffee. It was already a tradition for the prelaunch breakfast to be of steak. They emerged from breakfast still holding their bread, and found their old friend Joe Schmitt and his team of four technicians preparing their suits. Schmitt had suited up every American who had gone into space until then.

Following a carefully planned timetable they were wired up with their medical sensors and communication equipment, before slipping their long underwear on and spreading themselves on the special couches to suit up. First they put their legs

Figure 5.3. The Crew: a cheerful Apollo 11 crew before departure. Mission Commander Neil Armstrong (left), Command Module Pilot Michael Collins (centre) and Lunar Module Pilot Edwin Buzz Aldrin.

into the suits, pulled them up over their shoulders, and pushed their heads through the neck ring. Then they put on a light bubble helmet they liked to call the "Snoopy helmet", and were sealed off from the world. Their suits were then pumped up to a pressure of 131 kPa (27.6 kPa above sea level) of pure oxygen, checked for leaks, and for the next three hours they had to purge their blood streams of nitrogen so they would not get the bends. Then they walked out and climbed into the waiting white Transfer Van, its flashing light shooting red beams into the pale early morning light. Unable to hear the small crowd of people gathered to see them off, the three astronauts just grinned and waved back.

Meanwhile, up in the Command Module the back up Lunar Module pilot, Fred Haise, later to ride the ill-fated Apollo 13 mission, was working with Guenter Vendt, the Pad Leader and five other technicians, steadily and systematically going over the spacecraft, checking the 678 switches were in the right position. As his eyes and fingers flicked over the panels he could hear the familiar noises of the water glycol cooling system, cabin fans, and suit fans whirring away around him.

At 6:54 am while Armstrong grabbed the handrail and swung himself through the hatch of the Command Module and settled into the left couch, Collins paused on a narrow walkway from the elevator to savour the view and to consider the moment before climbing into the right couch.

Aldrin recalled, "While Mike and Neil were going through the complicated business of being strapped in and connected to the spacecraft's life support systems, I waited near the elevator on the floor below. I waited alone for fifteen minutes in a sort of limbo. As far as I could see there were people and cars lining the beaches and highways. The surf was just beginning to rise out of an azure blue ocean. I could see the massiveness of the Saturn V rocket below and the magnificent precision of the Apollo capsule above. I savoured the wait and marked the minutes in my mind as something I would always want to remember."[43]

Aldrin eased himself into the centre couch and waited for Joe Schmitt to couple him into the spacecraft. "My recollection of this procedure," he said, "was a mass of hands reaching and tugging from several directions. Mike, Neil and I were fairly helpless at this time – three totally battened down people waiting for the ride to start."[43]

With all the astronauts comfortable in their couches, the last task for Joe Schmitt before he left was to check the oxygen to the suits, leaving Fred Haise, the backup Lunar Module Pilot, to the final checks – straps, loose gear, and to look around for anything abnormal.

Haise was unable to speak directly to the recumbent astronauts, although he could hear them on the intercom. "They were locked up in their suits, and I went down in the capsule and assisted the two suit techs to transfer from the portable oxygen to the space-craft and strap them in. We all shook hands and as I crawled out I tapped Buzz on the shoulder. From the point of wishing them well this was the first attempt, so frankly at that time we didn't know they were going to land. All the missions have their chances for non successes, as I found out later in Apollo 13. We then exited and went through the hatch closure sequence and checked the leak test. Then we departed the pad to about the half way point, about an hour and a half before launch."

Unable to get up and look back and see the sunny world through the windows outside, the three astronauts settled down to wait for ignition, isolated from the Earth except through their intercoms. It was 7:52 am EDT.

The Launch

Light cumulus clouds topped by some cirrostratus moved in and the temperature drifted up to 29°C. A six-knot southerly breeze blew lazily in from the Atlantic. Across the swamps and reeds from the launch pad, thousands of cars, trailers, boats, caravans, tents, camper vans, and shelters of all kinds, were joined by crowds jamming every vantage point – waterways, roadsides, jetties, beaches – all trying to avoid the annoying mosquitoes while waiting anxiously for that brief moment von Braun described as, "The first time life will leave its planetary cradle, and the ultimate destiny of man will no longer be confined to these familiar continents we have known for so long."

At the official viewing site 5,000 special guests from all walks of life and from all over the world were there to witness the launch, along with 3,497 members of the press, many with direct lines to their offices around the world, such as in London, Paris, New York and Tokyo. On a beach someone had scrawled "Good Luck Apollo 11" in large letters in the sand.

The legendary American television commentator Walter Cronkite described his feelings of that moment as: "It was so much different from any other flight – it was something that had to grip you. You knew darned good and well that this was the real history in the making. The thing that made this one particularly gripping was that sense of history, that if this was successful this was a date that was going to be in all the history books for time evermore – everything else that has happened in our time is going to be an asterisk. I think we sensed that at the time – that this was it!"

The launch was the most hazardous part of the whole mission. If any part of the rocket, performing a superb balancing act, should touch the launch tower during the first twelve seconds, the entire Saturn V with its highly volatile fuels, could become a gigantic fireball. Jerry Lederer, NASA's Safety Chief at the time, calculated that the Apollo stack had 5.6 million parts, and 99.9% reliability still allowed 5,600 failures!

At the launch control center in Firing Room One 463 technicians and engineers, backed by 5,000 specialists on stand by, hunched over their consoles, intently studied their flickering screens, watching countless tests and checks of the whole vehicle being performed. When their task was completed, they announced *go* on the intercom loops.

Finally, all the checks were complete – the Saturn rocket was ready for its trip to the Moon. "Good luck and God speed from the launch crew," called Paul Donnelly, the Launch Operations Manager, on his microphone.

"Thank you very much," responded Armstrong.

The countdown started: "Twenty seconds and counting … fifteen seconds … guidance is internal … twelve … eleven … ten … nine. …"

The 750,000 people gathered to watch the launch were hung in suspense, listening to the count on their radios. All eyes were waiting for the first lick of flame at the base of the gleaming white rocket. Nearby, fourteen people in flame-protection gear on armoured personnel carriers tensed behind their sand bunker, ready to rush in to help the astronauts in an emergency.

At T-minus-8.9 seconds the five F-1 engines burst into life, spewing fire and smoke down the huge flame deflector below. 127,300 litres of cold water per minute flooded out over the walls, mixed with the searing flames to generate clouds of steam. Sheets of ice, formed on the rocket's skin from the super-cold fuel within, flaked off in an avalanche of white. Thundering shock waves spread out, filling the sky with startled flocks of duck, heron, and small birds. Even in the bright daylight the glare from the flames became so intense it hurt the eyes.

Four giant clamps gripped the straining rocket as the engines built up to their full thrust and the launch team rapidly checked all systems were running properly.

T-minus-zero, at 9:32 am EDT: "… all engines running. Lift off. We have lift off."

With maximum thrust built up to the equivalent of 180 million horsepower, equal to 32 Jumbo-jet aircraft, the hold down clamps released the straining rocket. At first slowly, majestically, the mighty vehicle rose off the pad, sliding out of eight guiding taper pins for the first 15 centimetres. The rushing river of searing flames plastered the gantry and created flecks of fire dancing on the steel structure. The fins at the bottom of the rocket cleared the tower and the flames and heat drew away to leave the blackened, blistered edifice standing empty, alone. 77,200 litres of water per minute still tumbled down around the base to preserve it from being destroyed by the heat from the rocket blast.

The huge crowd of onlookers stared through the heat haze at the shimmering image of the moonship and saw two gigantic torches of flame shoot out of the bottom and splay out to billowing clouds of fire and smoke. They watched in awe as the rocket began to rise off the ground in silence – it was eerie to watch this spectacle of fire and noise in silence.

But not for long. Fifteen seconds later it reached them. It began as the crackling of breaking sticks, rising to a barking, to a thunder louder than any thunderstorm they had ever heard. The very ground shook, the air pounded their bodies. A rushing wall of apocalyptic fury of sound engulfed them on its way to dissipate in the far distance. The ground vibrated up to six and a half kilometres away. The veteran Atlantic flier Charles Lindbergh, watching the launch, said it was like bombs falling nearby. Australian journalist Derryn Hinch, packed in among the press corps, said it was like being hit in the stomach with a cricket bat, and later he found bruises on the tops of his thighs from the shuddering desk top.

"The tower is clear," called the Launch Director. This call defined the moment that the responsibility for the mission was transferred from the Launch Control Center at the Cape in Florida to the Flight Director in the Mission Control Center at Houston.

Apollo 11 was on its way to rendezvous with the Moon, 350,980 suspense-filled kilometres away at that moment.

Up, Up – and Away

The three astronauts, firmly strapped to their couches above the thundering rockets, were not aware of the actual moment of lift off, but first felt a powerful, insistent thrust to their backs accompanied by a distant rumble rather like the sound of an express train. Slowly, then faster and faster the rocket climbed, the three men pressed irresistibly into their carefully contoured couches as they felt the surging power of those five engines flowing through their bodies, through everything around them. During the first fifteen seconds or so they were thrown left and right against their straps in spasmodic little jerks as the huge vehicle adjusted itself to the course and wind effects, then speared up into the brilliant blue sky. In about 40 seconds the 36-storey-high vehicle was travelling faster than the speed of sound, and the noise in the cabin dropped away. As they slowly built up to a weight four times greater than on Earth it needed a strenuous effort for the astronauts to raise their arms to reach for a switch.

The swampy coast of Florida, beginning to bake in the hot morning sun, rapidly receded behind the trail of white smoke as the mighty Saturn V motors gobbled up 2,128 tonnes of fuel at the rate of 13.3 tonnes per second, to drive the 3,198 tonne monster from rest to 10,140 kilometres per hour in an incredible two-and-a-half minutes. The massive vehicle lost more than three quarters of its weight in the first 160 seconds of flight. As the heavy vehicle rose, consuming its fuel at this incredible rate, it rapidly became lighter, as it became lighter it could go faster, as it went higher and faster it met less air resistance, and went still faster until it was travelling nine times the speed of sound in those first awesome 160 seconds.

Seventy-two kilometres above the crowded swamps, the massive first stage, now empty, dropped off to head for the Atlantic. The power of the Saturn V rocket was so great that the whole vehicle shrank while accelerating, then when the engines ran out of fuel and the thrust momentarily stopped, it snapped back to its normal length and jolted the astronauts forward before the second stage fired and continued to push them on to a speed of 24,140 kilometres per hour. At a height of 97 kilometres the escape tower and cover were blown off the Command Module, and for the first time the astronauts could see ahead out of the windows. As they streaked through the thinning atmosphere, the blue sky darkened to the black of space.

At a height of 177 kilometres, and 2,253 kilometres down range from the launch pad, the third stage took over and with a lot of noise and motion, shoved them into Earth orbit. It only took them 12 minutes to go from sitting motionless on the launch pad to be hurtling along in orbit at a speed of 28,162 kilometres per hour, 191 kilometres above the Earth. Apollo 11 was officially recorded in the US Defense Department's log as "Man-made object in space No. 4039".

Once in Earth orbit the astronauts enjoyed the experience of weightlessness while watching the continents, blue seas, and clouds sliding past the windows. On the Earth below 17 tracking stations, 4 ships, and 8 Boeing 707 jet aircraft followed every move by the spacecraft. Apollo 11 was flying along upside down so the base of the Command Module could point to the black sky and stars to allow Collins to take star sights with the sextant. As they went over the Canary Islands the astronauts removed their helmet and gloves, and settled down to check out the spacecraft was ready for the big voyage.

They raced across the Indian Ocean towards Carnarvon in Western Australia, the first station to send information to Houston confirming that Apollo 11 was in its proper Earth orbit. All the Carnarvon antennas were locked firmly on the horizon – waiting.

Paul Oats, Deputy Director at Carnarvon, recounted, "We had three tracking systems operating. When it came up over the horizon the Americans were desperate to run their checkout programmes and make sure everything was working properly. We were locked up solid on the spacecraft before line of sight about 4,828 kilometres away over the Indian Ocean, and then we tracked it over towards the eastern coast of Australia."

Apollo 11 rushed on through the night, and after a brief glimpse of the Goldstone station in California where they managed to snatch a minute's worth of television to the ground, they prepared for firing the Saturn rocket to kick them onto the track for the Moon. Their orbit had been very accurately measured through the tracking stations, all checks were completed, and now they were positioned ready to leave the Earth. Calculations figured a burn of 5 minutes 47 seconds at 2 hours 44 minutes after liftoff over the Pacific Ocean. In the darkness below them, Boeing 707 Apollo Range Instrumented Aircraft (ARIA's), were following their track to record every move they made for transmitting back through tracking stations to Houston later. Following mission rules, the astronauts put their helmet and gloves back on.

It's *Go* for the Moon

The second orbit over Carnarvon was the big moment for the Capcom at Houston to tell the astronauts they were to go ahead for the lunar burn.

> McCandless: "Apollo 11, this is Houston. Slightly less than one minute to ignition and everything is *go* – slightly less than one minute to ignition."
> Armstrong: "Roger … Ignition."

Figure 5.4. Legendary American television commentator Walter Cronkite visited Goldstone on the 4th of July to absorb the atmosphere of a tracking station before Apollo 11 was launched. Here he is discussing the operation of the antenna with Dee Yeaman at the Servo Console.

The astronauts relied on the spacecraft computer automatically selecting the attitude and controlling the firing of the rocket motor for the burn at 12:22:13 EDT. The rocket fired right on time. As the astronauts were gently pressed into their couches with a pressure of about 1 g, they felt a surge of relief tempered with excited tension.

The spacecraft began to break away from Earth orbit and rapidly gained altitude as it headed east towards the dawn. A few moments later the Sun was shining into the windows and, right on time, the engine shut down. Armstrong relayed back, "Hey, Houston – Apollo 11. That Saturn gave us a magnificent ride."

They were on track to the Moon, travelling at 39,000 kilometres per hour.

The first task was to turn the CSM around and dock with the Lunar Module, still nestling in the end of the Saturn rocket. As Collins was the pilot for the manoeuvre, the three astronauts had to change places before he could take control.

As Collins brought the two vehicles gently together, he noticed the Lunar Module's flimsy aluminium skin was so thin it rippled to the bursts of gas from the Command Module's control jets like a breeze across long grass. The two spacecraft mated together, the twelve latches clamped tight, and Collins clambered down to remove the hatch before inspecting the tunnel and clamps, and piloted the CSM away from the Saturn rocket, now looking rather forlorn and empty. They went separate ways, the Saturn going off into solar orbit, and the CSM now locked with the Lunar Module heading for the Moon.

Before lunch they helped unzip each other's pressure suits and stowed the bulky garments out of the way under the centre couch. Dressed in white nylon jump suits they were much more comfortable now, with more space in the cabin The three astronauts settled down to lunch: a meal of beef and potatoes, butterscotch pudding, four brownies (thin, dense, chocolate cakes), washed down with grape punch.

At 30,577 kilometres from Earth Aldrin looked out the window, and was surprised to see the whole Earth framed in the window for the first time. From now on the spacecraft was in full sunlight all the time – looking towards the Sun it was blinding, while looking away it was jet black. The stars were invisible because of the light reflecting off the spacecraft. The next task was to initiate a steady roll to keep an even temperature all over the spacecraft, and set a roll rate of one turn every twenty minutes. The Moon and the Earth appeared alternately through the windows as the spacecraft rolled along towards its lunar rendezvous.

During the first day the Moon didn't seem to be getting bigger, but the Earth was visibly shrinking. Looking out of the windows, they noticed that the Earth looked very bright, the blue oceans and white clouds standing out, but found the green jungles harder to spot.

They had no night now, their sleep periods and "day" were determined by the time back in Houston. To sleep they covered all the windows and the two off-duty astronauts climbed into their sleeping bags. Remembering the tight confines of the Gemini spacecraft it was a strange sensation to be floating in darkness with no body weight pressing down on a mattress. Although he was really weightless, Collins felt he was lying on his back, not his stomach, purely because the instrument panel was in front of him. On Earth he was conditioned to think of the panel as "up". They turned the radio down, dimmed the lights, and drifted off to sleep listening to the buzzing sound of electric fans and the occasional thump of the attitude controlling thrusters. The radio went silent – the night shift at Houston would only disturb them if there was an emergency.

The days passed quietly, with Houston keeping them informed of selected events back on the Earth, and putting on TV shows that were beamed around the world.

During one of these shows Collins told the people on Earth: "We do have a happy home. There's plenty of room for the three of us and I think we're all willing to find our favorite little corner to sit in. Zero-g's very comfortable, but after a while you get to the point where you sort of get tired of rattling around and banging off the ceiling and the floor and the side, so you tend to find a little corner somewhere and put your knees up, or something like that to wedge yourself in, and that seems more at home."

He also found that weightlessness changed the character of the spacecraft. On Earth the tunnel into the Lunar Module was an inaccessible space overhead, but on the lunar journey it was a cosy corner to wedge yourself in without the need for a restraining belt.

At the Tidbinbilla Tracking Station near Canberra the staff were all settled down and looking forward to their key role tracking the Lunar Module on the Moon's surface. At 6:25 pm local time on 18 July there was a fire in the power input circuit breaker of the backup transmitter. Looking at the damage, they first estimated it would be at least a week's work to repair it. But there wasn't a week left.

Alan Blake, the Transmitter Field Engineer, described the events: "I had just come down from dinner in the canteen when there was a call on the intercom to say there was a fire in the transmitter power supply. I ran out there, and found dense, thick black smoke was everywhere, and knocked off the circuit breaker and waited for the smoke to clear, then went into the cabinet. There was a horrible mess in there. The temperature had been so hot that the top of the cabinet had buckled. It took quite some time for it to cool down enough to touch it.

"The only thing to do was to jury rig the thing to get it back on air. We chopped out all the old chunks of cable loom and made up new bunches of wires. In the meantime we pulled out the removable units and there was a queue of blokes from the other subsystems. We gave them the units and drawings and they went down to the store and used anything and everything that could fit in. Ken Cox and I worked through the night, and we had arranged for Geoff Rose to take over the next day. We were down for no more than about 12 hours."

Apollo 11 Arrives at the Moon

On the fourth day, Apollo 11 arrived and Collins swung *Columbia* around to look at the Moon. Not having seen the it for a day, all three astronauts were startled by the dramatic change in its appearance. Now it looked huge – filling their windows with its three-dimensional bulging midriff, so close they felt they could reach out and touch it. Bathed in subtle earthshine and ringed by sunlight spilling around the Moon's rim, the dark crater-studded surface below them looked spectacular – and full of mystery.

Armstrong reported, "The view of the Moon that we've been having recently is really spectacular. It fills about three quarters of the hatch window and, of course, we can see the entire circumference, even though part of it is in complete shadow and part of it is in earthshine. Its a view worth the price of the trip."

Just as they were finishing breakfast the windows darkened as they entered the shadow of the Moon and all the stars came out: "Houston, its been a real change for us. Now we are able to see stars again for the first time on the trip. The sky is full of stars, just like the nights out on Earth."

Into Lunar Orbit

Shortly before the spacecraft was due to go behind the Moon for the first time, and out of sight of the Earth, Houston announced, "Eleven, this is Houston. You are *go* for LOI, over." This meant that they could burn their rocket motor to put them into lunar orbit, LOI being "Lunar Orbit Insertion". The burn to slow them into orbit and not shoot right past had to take place behind the Moon, out of sight of Earth and the tracking stations.

Aldrin responded, "Roger, *go* for LOI."

At 2:13 pm on Monday 19 July, the spacecraft disappeared behind the Moon travelling at a speed of 8,408 kilometres per hour. The astronauts checked and rechecked the procedure for the burn for what seemed hundreds of times as just one erroneous digit entered in the computer could turn them around and blast them into orbit around the Sun instead of the Moon. They anxiously scanned their displays to keep an eye on the progress of the six-minute burn to put them into a lunar orbit of 98 by 272 kilometres. It performed flawlessly.[33]

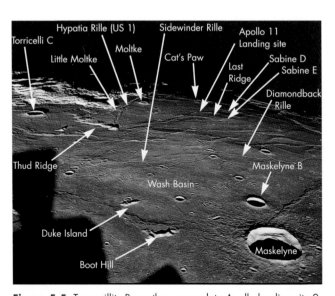

With no signal from the spacecraft Mission Control in Houston went quiet, the subdued Flight Controllers mainly seated at their consoles, some standing up, an odd conversation taking place. Bill Anders and Jim Lovell, two of the Apollo 11 back up crew arrived and joined astronaut McCandless at the Capcom console. They were all waiting for the spacecraft to reappear to confirm the burn had gone all right. The tracking station at Madrid found the signal right on time, the astronauts busy aligning their antenna for the

Figure 5.5. Tranquillity Base: the approach to Apollo landing site 2, in the southwestern corner of the Sea of Tranquillity, taken from orbit while *Eagle* was still docked to *Columbia*.

best angle to Earth as they came over the lunar horizon. Houston were anxious to know how the burn to put them in lunar orbit had gone.

> McCandless at Houston: "Apollo 11, this is Houston. How do you read?"
>
> Collins: "Read you loud and clear, Houston."
>
> McCandless: "Could you repeat your burn status report?"
>
> Collins: "It was like … it was like perfect."

A second, critical burn of seventeen seconds put them into an elliptical orbit of 87 by 106 kilometres, ready for the Lunar Module to depart for the lunar surface. Critical, because even a burn two seconds too long could put them on a collision course with the other side of the Moon.

As they approached the Sea of Tranquillity they could see it was early dawn on the surface below them – the Sun was just tipping the peaks and boulders with a rich glow while long, jumbled, black shadows stretched across the crater studded surface. Armstrong and Aldrin were riveted to their windows trying to make out the landmarks for the next day's landing, but they were disappointed to find the actual landing site seemed to be in darkness, so they were not able to preview the landing spot.

> Armstrong reported his observations: "We are currently going over Maskelyne … and Boot Hill, Duke Island, Sidewinder. Looking at Maskelyne W – that's the yaw around checkpoint, and just coming into the terminator – at the terminator it's ashen gray. If you get further away from the terminator, it gets to be a lighter gray, and as you get closer to the subsolar point, you can definitely see browns and tans on the ground."
>
> McCandless: "Roger, Eleven. We are recording your comments, for posterity."
>
> Armstrong: "… and the landing site is well into the dark. I don't think we're going to be able to see anything of the landing site this early."

Collins was disappointed with the view of the lunar surface out of his window. He felt that compared to the variety and exciting colours of the Earth from space this monotonous plaster-like surface peppered with endless craters of all sizes looked distinctly unwelcoming.

It was planned to take three hours to prepare the Lunar Module for the landing. On the fourth "night" they were to sleep in lunar orbit. Before covering the windows and turning down the lights they carefully prepared all the clothing and equipment they were going to use the next day. With thoughts of the momentous events to come the next day, they had some difficulty getting to sleep.

But the day did come. Evans at Houston called up, "Apollo 11, Apollo 11 … good morning from the Black Team."

The insistent voice finally penetrated into Collins' consciousness, rousing him from a deep, deep sleep. After a few moments his groggy voice answered, "Good morning, Houston."

> Evans: "Good morning. Got about two minutes to LOS here, Mike."
>
> Collins: "Oh my, you guys wake up early!"
>
> Evans: "Yes, you're about two minutes early on the wake up. Looks like you were really sawing them away."
>
> Collins: "You're right. How are all the CSM's systems working?"

Evans: "Eleven, Houston. Looks like the Command Module's in good shape. Black team has been watching real closely for you."

Collins: "We sure appreciate that because I sure haven't."

Eagle Leaves Michael Collins in *Columbia*

After sipping lukewarm coffee, and chewing a breakfast of bacon cubes, the astronauts dressed in their spacesuits again, and at 10:27 am Aldrin crawled into the Lunar Module and switched on all the systems ready for their departure. Looking around he saw the cabin, "… as charming as the cab of a diesel locomotive. Weight restrictions prevented the use of panelling, so all the wiring bundles were exposed. Everywhere I looked there were rivets and circuit breakers. Even the safety covers had been removed from the circuit breakers and switches. The hull had been sprayed with a dull gray fire-resistant coating."

An hour later Armstrong joined him, and they transferred the last items from the Command Module before they closed and bolted the hatches.

At 8:00 am on Sunday morning of 20 July, in the Mission Control Center, Houston, the White Team drifted into their positions led by the dynamic Eugene F. Kranz, dressed in his traditional white brocade waistcoat made for him by his wife, Marta, and fortified with a fix of his favourite Sousa marches. This was the team specially trained for the landing. Kranz thrived on pressure and fast decision making action. He could conduct a number of conversations simultaneously on the intercom loops, and could recall up to 2,500 emergency procedures from memory, "I don't waste even a few seconds looking up the right solution in a book where a couple of seconds may make all the difference," he explained.

Astronaut Charlie Duke seated himself at the Capcom's position, the only person normally in direct communication with the men on the Moon. Armstrong had specifically requested Duke to be Capcom for the landing because of his intimate knowledge of the Lunar Module systems.

In the viewing room most of NASA's elite were gathering to witness history, history largely created by their own tireless dedication to the project – NASA Administrator Thomas Paine, rocketeer and Director of the Marshall Space Flight Center Wernher von Braun, Director of the Kennedy Space Center Kurt Debus, Apollo Program Director four star General Sam Phillips, Associate Administrator for Manned Space Flight George Mueller, plus most of the astronauts.

Flight Director Gene Kranz directed that the doors be locked, and from that moment the suspense began to grip the people watching the controllers bent over their consoles. Were the doors locked to keep people out, or the flight controllers in?

In the spacecraft, after a long and tedious checkout procedure, which included deploying the Lunar Module's landing gear, Collins threw the switch that released the Lunar Module at 2:46 pm, and watched them through the window as Armstrong slowly turned the Lunar Module around for Collins to inspect the spacecraft.

They were behind the Moon out of sight of the Madrid tracking station in Spain. When they reappeared, there would be two spacecraft – the Command and Service Module *Columbia* and the Lunar Module *Eagle*. The Goldstone tracking station in California would come into view of the Moon before the Lunar Module reached the surface. Honeysuckle Creek was in darkness right out of sight on the other side of the world. Tension began to spread insidiously through Mission Control and the tracking network as the clocks around the world and on the Moon relentlessly swept away those final moments in the flight plan everyone in the mission was following – step by step.

> Aldrin: "The *Eagle* has wings!"
> Collins: "I think you've got a fine looking flying machine there, *Eagle*, despite the fact that you're upside down."
> Armstrong: "Somebody's upside down!"
> Collins: "OK *Eagle*, one minute … you guys take care."
> Armstrong: "See you later."

Collins watched the *Eagle* dwindle into a dot in the distance. He was now alone; if anything happened to his mates he would be the most isolated person in the history of mankind. He could not help considering the odds and thinking about all the events, the unknown dangers the two were about to face, and felt it was a fifty–fifty chance he could be alone for the rest of the trip back to Earth. Armstrong figured out the odds about the same. Now it was Collins' job to keep *Columbia* ready for their return – if they did return.[33]

In Houston the White Team had now settled into their routine, Kranz going around a final call:

> Kranz: "Got us locked up there, TelCom?"
> Puddy: "OK, it's just real weak, Flight."
> Kranz: "OK, how ya lookin'? All your systems *go*?"
> Puddy: "That's affirm, Flight."
> Kranz: "How about you Control?"
> Carlton: "We look good."
> Kranz: "Guidance, you happy?"
> Bales: "*Go* with systems."
> Kranz: "FIDO, how about you?"
> Greene: "We're *go*. We're a little low, Flight, no problem."
> Kranz: "Rog."
> Kranz: "OK, all flight controllers, thirty seconds to ignition."

The Lunar Landing

Duke called up, "Eagle, Houston. You are *go*. Take it all at four minutes. You are *go* to continue powered descent."

The two astronauts felt a thrill of anticipation and grinned at each other through their helmet visors. Armstrong nodded his head: "Roger."

Altitude 12,192 metres. Aldrin took his eyes off the control panel for a moment to look through the window. "And we got the Earth straight out our front window."

They headed down for the Moon's surface, looking up at the Earth through the windows, unable to see the Moon beneath them. At all times they were very conscious of their home: the Earth – a blue and white jewel glittering in the black void of space – hanging suspended up there in the lunar sky. The *Eagle* must have felt very remote and alone in that hostile environment. The land they were heading for had no water or food or shelter, and the nearest friendly mechanic or technician and their spare parts were three days and over 321,860 kilometres away across the void.

The Lunar Module had begun racing horizontally across the lunar landscape at 5,793 kilometres per hour when it left the CSM, and now had to slow up in stages, to the speed of a jet plane, 965 kilometres per hour, down to the speed of a car at 100 kilometres per hour at an altitude of 2,133 metres, finally hovering above the surface, before dropping vertically into the dust. Armstrong and Aldrin had to make the first ever landing on the Moon in one go – there was no chance of a second attempt.

Altitude: 10,211 metres. Then, quite unexpectedly, a yellow caution light winked at the astronauts from the computer control panel. It was identified as a 1202 alarm. They automatically asked the computer to define the problem. I am overloaded, it answered in its code, *I can't handle all the jobs you're giving me in the time available.*

Armstrong warned the flight controllers at Houston. "Program alarm."

Duke replied, "It's looking good to us, over."

Armstrong persisted. "It's a 1202."

Capcom Charlie Duke said of that moment, "When I heard Neil say '1202' for the first time, I tell you my heart hit the floor. I looked across at Steve Bales but he was busy at his console and came back with the answer almost straight away – we were *go*."

Jack Garman, a young back-room expert supporting Bales from another console, remembered a similar problem had been tried out in a simulation only a week or so before, quickly reassured Bales: "Its executive overflow; if it does not occur again, we're fine."

Twenty-six-year-old Steve Bales recalled that fateful moment: "I had just started to relax a little bit, if you can call it relaxing, and I heard the program alarm, and quite frankly, Jack, who had these things memorised said, 'that's OK', before I could even remember which group it was in. … I was frantically trying to look down … by the time I looked at the group and saw which one the alarm was in, Jack said it's OK, I remembered – yeah, that's one of those we said it's OK, I looked up, the rest of the computer looked good, so I said 'Lets go!' It took us a long time. In the Control Center any more than three seconds on descent is too long … and it took us about ten to fifteen seconds."

Swiftly dropping down to the Moon's surface, both astronauts confirmed the location of the "ABORT" button and sweated out a thirty second pause while at Mission Control Kranz snapped out a final tense roll call around his flight controllers. Steve Bales' decision alone decided the fate of the mission, to abort and terminate the mission then and there, or continue on to success or … the possibility of a disaster. As it turned out it was the right decision, and Bales later collected his Medal of Freedom from the President along with the astronauts.

In the middle of the 1202 crisis, Chuck Deiterich in Retro chopped in: "Flight, Retro."

Kranz: "Go Retro."

Deiterich: "Throttle down six plus twenty five."

Kranz to Duke: "Six plus twenty five."

Retro was advising Kranz to pass on to Duke that 6 minutes 25 seconds into the burn the crew should expect the engine to throttle down to 55 per cent power.

Duke: "Roger. We got you. We're go on that alarm … six plus twenty five throttle down."

Armstrong: "Throttle down on time."

Aldrin: "Throttle down on time! You can feel it here when it throttles down. Better than the simulator."

Prepared by the endless simulations and intense training every person involved worked as a tight team. Director of Flight Operations Chris Kraft chose his flight controllers, their average age 26 years, very carefully: "They need nerve and an intangible ingredient, a sort of creative intuition in arriving at good answers. I'm not talking about blind hunches, but there is an element of black magic in the kind of human judgment that has to be stirred into this computerised recipe for controlling a mission." These elements were needed as split-second, life or death decisions came rapidly as both height and fuel began to run out, while beneath the Lunar Module the landing area began to take shape.

Duke kept up the flow of information. "You're looking great at 8 minutes."

Altitude: 2804 metres. Now the Lunar Module began to drift up from its horizontal attitude, slowly its legs began to point down to the Moon's surface, and the astronauts could begin to see the Moon's surface in the bottom of their windows.

Armstrong was trained to land the Lunar Module. The two pilots had to work together as a cohesive team, Armstrong controlling the spacecraft's flight while looking out of the window at the landing site; Aldrin concentrating on the display panel and calling out the information he needed. Armstrong had to translate what he saw with what he heard with what he felt to the spacecraft controls to guide the *Eagle* safely down to the lunar surface.

> Armstrong: "OK. Five thousand (feet). One hundred feet per second is good. Going to check my attitude control. Attitude control is good." (The hundred feet per second is the descent rate.)
>
> Duke: "Roger. Copy. Eagle, Houston. You're *go* for landing. Over."
>
> Armstrong: "OK. Three thousand at seventy."

At 610 metres above the lunar surface another alarm winked from the computer. Aldrin called out, "1201," with growing concern. With no time for explanations from Houston, they had to trust their lives to the judgment of the flight controllers.

Armstrong repeated, "1201!" Then, "OK, 2000 at 50."

Duke acknowledged, "Roger, 1201."

In Mission Control Kranz queried Bales again, "1201 alarm?"

Bales had already been on to Garman, "Same type, we're *go*, Flight."

> Kranz to Duke: "OK, we're *go*."
>
> Duke: "1201 alarm. We're *go*. Same type. We're *go*."
>
> Aldrin: "2000 feet. 2000 feet"
>
> Duke: "Eagle looking great. You're *go*. Roger 1202. We copy it."

Armstrong was riveted to his controls: "Now we get to that final landing phase and this is altitude versus range to the landing site. This is about the last 3/4 mile into the touch down spot from a thousand feet (305 metres). This part is normally flown automatically and as you get down to 500 feet (152 metres) you have some options as to what you can do to complete the landing. One is to just leave the thing run automatically. Then there's several manual options that you can choose from. One is manual attitude control but with an automatic throttle that will control the descent rate to the programmed value that it thinks it should have. One is manual attitude control with a rate of descent mode on the throttle so that you can actually command your descent rate and it'll freeze. Say you're coming in at 17 feet (5 metres) per second, it'll hold 17 feet per second down until you put a blip on the switch and each blip changes your rate of descent mode by one foot (0.3 metre) per second. I really didn't think that was likely to work, but it did. Matter of fact, it was quite smooth.

"The final method that you have is manual attitude and manual throttle. Just hand on throttle like most of our rudimentary VTOL (Vertical Take Off and Landing) aircraft and like you would fly a helicopter. Now, you could fly auto, but its not likely that many test pilots would do that. One reason is that the auto system doesn't know how to pick a good area and can't change its mind. The second is that when you get right down to the final phases and it turns out there is a little residual velocity of a couple of feet per second sideways – you'd have a bad case of stubbing your toe on touch down. For those reasons, I didn't intend to make an automatic landing; it was my intention to fly the manual mode with this one foot per second incremental rate of descent mode on the throttle into touchdown, which is what I did.

"But as we got to the point where you'd normally take over manually, I had been looking out the window and, if you had been listening at the time, all we really saw was a gigantic crater and lots of very big rocks – a very unfavourable position to land. Now it looked like we might be able to land short and I was really tempted for a minute because I knew the scientists would have a ball if we could land in the middle of that boulder patch. They would think it was just Jim-Dandy if we could run up on the rim and take pictures down the sides of this really big crater and be overjoyed; and I thought about that for a little bit – and I didn't do it. It's an old rule, when in doubt, land long, and I did. We extended the range down about 1,100 feet (335 metres) past where it would have gone if we had let it go automatically.

"I didn't have any of those 30 storey rocks that Tom (Stafford) looked at, but I thought that this area with all those automobile sized rocks wasn't probably a good place for me to try and join them. Well, I thought this was a good spot and then I got closer and decided it wasn't so I changed the descent rate and changed the attitude and went on a little bit further and thought this was a good spot, and when I got closer, I was dissatisfied and was just absolutely adamant about my God given right to be wishy-washy about where I was going to land."[44]

Back in the Mission Control Center in Houston the flight controllers were quiet, there was little they could do now, but they were getting jittery – why wasn't Armstrong landing? He should have landed by now – he always had in the simulations. There were no clues coming down the voice channel – just figures from Aldrin. They were all staring at their consoles, helpless, not one of them knew why the *Eagle* was still weaving about above the surface, but all were acutely aware that time and fuel were fast running out. What was going on up there?

At a height of 76 metres Aldrin flicked a glance out of his window and had a fleeting impression of the Lunar Module's shadow on the lunar surface ringed with a halo of bright sunlight before a red warning light came on – only 5 per cent fuel remained – and they still weren't down. There were only 94 seconds left to land. Kranz remembers, "That really grabbed my attention, mainly because during the process of training runs we had generally landed by this time. Now it was a question of continuing the countdown. It was a horse race between running out of fuel or getting down on the surface."

Aldrin recounted, "At an altitude of fifty feet (15 m) we entered what was accurately referred to as the dead-man zone. In this zone, if anything had gone wrong – if, for example, the engine had failed – it would probably have been too late to do anything about it before we impacted with the Moon. There were no fail-safe abort systems available until landing. I felt no apprehension at all during this short time. Rather, I felt a kind of arrogance – an arrogance inspired by knowing that so many people had worked on this landing, people possessing the greatest scientific talents in the world."[43]

From out of the airless black sky above the paste-white lunar surface bathed in the contrasty early morning sunlight, the Lunar Module appeared with a roaring rocket motor blasting a stream of gases down at the surface. Life from Earth had arrived at the Moon and brought their inevitable pollution, confusion, and debris.

Like a prehistoric predator, its two windows like beady eyes above the four dangling legs, the Lunar Module now hovered 9 meters above the surface, instruments and astronauts desperately searching, trying to probe the lunar dust for a clear spot to land. Brilliant searchlights sent piercing shafts of light through the lunar dust. Armstrong could see small boulders and rocks sticking up out of the blanket of dust blasting away from their rocket motor. A hard white surface appeared through the dust, followed by black shadows of the approaching legs and spindly probes.

Inside the Lunar Module, isolated from the rocket blast and dust outside, Aldrin was busy reciting facts and events displayed on the console in front of him: ". . six … forward … lights on … down two-and-a-half … forty feet … down two-and-a-half … kicking up some dust … thirty feet … two-and-a-half down … faint shadow … four forward … four forward … drifting to the right a little … OK. …"

The tension was becoming unbearable.

Aldrin continued, "Four forward. Four forward. Drifting to the right a little. Twenty feet … down a half."

Duke reminded them, "Thirty seconds." (i.e. fuel remaining)

Kranz recalls, "We escalated another notch when we got the 30 second call. The next thing we would start doing would be to call down every second from 15 seconds on down the line."

Armstrong's version is: "Well, they called 60 seconds from the ground, and they called 30 seconds, and I heard that, and the next thing I was supposed to see was the contact light but I never did see that – that blue light. They tell me it did come on and Buzz saw it and he called it, but I never saw it. I was all eyeballs out the window at that point. You know we had some problem with dust – the exhaust kicking up dust – and it obscured the surface and made it a little difficult because it was flying off parallel to the ground at a very shallow angle and at very high speed, like ground fog. You could see through it, you could see craters and rocks through it and if you had been expecting it, and I should have been, we probably would have neglected it. I'm sure the next crews won't have that kind of problem. In fact, it did confuse us a little bit. Although it didn't affect the altitude determination very much, I did have trouble figuring out what my cross range and down range velocities were and I didn't want to stub my toe on touch down.

"We were supposed to take over about 3:30 in the count down and get a low level light, which occurs when you have about 5% of fuel remaining and touch down right about the same time. Well, we took over just a little bit late and got the low level light on time – I saw that. That gives you about 94 seconds of flying time left at that point. You have to save the last 20 seconds for an abort. We're flying in a dead man's curve down here close to the ground. If the descent engine quits, the ascent engine is unable to be ignited to go through its ignition sequence and get you back on a safe abort before you hit the surface. So, we were, of course, saving those last 20 seconds so that if we did need to abort we could 'hang the chili to it' as they say in Texas and get out of there while we still had the big powerful descent engine. Then when we ran out of fuel, we could stage, and have plenty of time to get going with the smaller 3,500 lb thrust ascent engine.

"Now I deviated from the plan here a little bit. Our idea was that we were going to get to 5 feet (1.5 metres) and let those probes – the ones sticking out the bottom of the Lunar Module's legs – touch the ground. They light a blue light in the panel. Then I was going to go about another second which would get me down to about 3 feet, say I was coming down about 2 feet per second, and then I'd punch the stop button. Now its been against my grain to shut off the engine when I was in the air, but it was supposedly an important thing to do because it would prevent the engine from blowing up as it got very close to the surface, or it would avoid overheating of the bottom of the Lunar Module. Also if we hit hard enough, we would collapse those struts so that the stairsteps on the front would be close enough to the surface so we could get comfortably down.

"Well, I forgot all that when I got down and actually touched down at a very low velocity, very much like what you'd be used to in a normal helicopter landing. Turned out the thermal effects weren't so bad and the engine didn't have any problem and it *was* a long way from the top stair down to the surface, but we were able to make that $3\frac{1}{2}$ feet (1 metre) or so."[44]

In a maelstrom of dust, lights, shadows, legs, and spent gases, the spaceship *Eagle* from Earth touched gently down on the lunar surface at 3:17:39 pm on 20 July (6:17:39 am AEST, 21 July 1969.)

Aldrin: "Contact light!."

Armstrong: "Shutdown."

Aldrin: "Okay. Engine stop."

Aldrin remembers, "At ten seconds we touched down on the lunar surface. The landing was so smooth I had to check the landing lights from the touchdown sensors to make sure the slight bump I felt was indeed the landing. It was."[43]

Duke confirmed, "We copy you down, *Eagle.*"

The billowing dust just dropped and all was still. Suddenly all the gut-wrenching, urgent decisions were gone – just silence. With no atmosphere there were no familiar sounds from outside, no rustling leaves, no bird calls or human or animal noises, just the sound of their own breathing inside their helmets.

The first human voices on the Moon crackled over the intercom and were relayed to the 600 million earthlings holding their breath. As they all heard the first words from another world in English with an American accent, it seemed that for the first time in history the human inhabitants of the Planet Earth were globally united. This strange creature from Earth, the Lunar Module called *Eagle* was safely down on the lunar surface in an area ringed on one side by fairly good sized craters, and on the other side by a boulder field, about the size of a house lot.

Armstrong and Aldrin looked at each other through their visors, reached across and vigorously shook gloved hands, excited by the tension of the events on the way down, before Aldrin responded automatically to their training procedures and began to prepare for an emergency launch, when he was surprised to hear Armstrong announce: "Engine arm is off. ... Houston, Tranquillity Base here. The *Eagle* has landed!"

Duke: "Roger, Tranquillity. We copy you on the ground. You've got a bunch of guys about to turn blue. We're breathing again. Thanks a lot."

Aldrin: "Thank you."

Duke: "You're looking good here."

Armstrong: "OK. Let's get on with it. OK, we're going to be busy for a minute."

Charlie Duke gratefully sank back into his chair, took a deep breath, and exchanged grins with Deke Slayton. He could hardly believe it had happened.

"OK, everybody – T1, stand by for T1," Kranz rasped out to the flight controllers while Duke was still saying, "We copy you on the ground", but then for a moment he was speechless. The 35-year-old crew-cut Kranz, who had the flight control team and himself under rigid control all the way down, admitted, "On the consoles for the TV tubes they've got two handles and I found myself with my left hand holding onto that handle like the console was going to run away and I kept scribing my notes and the paper kept rolling up on me because I'd be embedding notes that I was taking during descent, and when we finally got down on the surface the viewing room … there were no people in there during training … they started cheering … that's when I finally found, my God, we'd landed!

"When the viewing room erupted, I sorta froze and was speechless and just rapped my arm on the console and broke my pencil and bruised myself from my palm all the way up to my elbow."

It was enough for him to regain control and coolly announce, "All right, everybody settle down, and lets get ready for a T+1 STAY/NO STAY."

> Duke: "Roger, Eagle, and you are Stay for T1. Over. Eagle you are Stay for T1."
>
> Armstrong: "Roger, understand – Stay for T1."

T+1 was one minute after landing – decision time for staying or launching in a hurry if there was danger to the astronauts or spacecraft. There were only three minute or twelve minute abort points – after twelve minutes they would have to wait for Collins in *Columbia* to go around the Moon again.

> Duke: "Rog., Tranquillity. Be advised there are lots of smiling faces in this room, and all over the world."
>
> Armstrong: "Well, there are two of them up here."
>
> Duke: "Rog. That was a beautiful job, you guys."
>
> Collins from 96 kilometres above: "And don't forget the one in the Command Module."
>
> Duke: "Tranquillity, Houston. We have you pitched up about $4\frac{1}{2}°$. Over."
>
> Armstrong: "That's confirmed by our local observation."
>
> Duke: "Roger."

While Armstrong and Aldrin were in constant communication with Mission Control, Collins in the Command Module was spinning around and around and around the Moon, relying on somebody relaying the events to tell him what was happening. After forty minutes of complete isolation behind the Moon on each orbit, he could talk and listen to the Earth for seventy five minutes through Goldstone, and later Tidbinbilla and Madrid, but he only had about seven minutes in touch with *Eagle* each time he passed over Tranquillity Base. Then it was back to another forty minutes of isolation. He was desperately hungry for news on what was happening to his mates on the surface.

> Duke: "Rog. Columbia this is Houston. Say something – they ought to be able to hear you. Over."
>
> Collins: "Roger, Tranquillity Base. It sure sounded great from up here. You guys did a fantastic job."
>
> Armstrong: "Thank you. Just keep that orbiting base ready for us up there now."
>
> Collins: "Will do."

"If there was any emotional reaction to the lunar landing," Aldrin said later, "it was so quickly suppressed that I have no recollection of it. We had so much to do, and so little time in which to do it, that no sooner had we landed than we were preparing to leave in the event of an emergency. I'm surprised, in retrospect, that we even took time to slap each other on the shoulders."[43]

Later, at the tenth anniversary celebrations, Armstrong confessed, "If there was an emotional high-point, it was the point after touchdown when Buzz and I shook hands without saying a word. That still in my mind is the high-point."

Aldrin explained, "I stared out at the rocks and shadows of the Moon. It was as stark as I'd ever imagined it. A mile away, the horizon curved into blackness. It was strange to be suddenly stationary. Spaceflight had always meant movement to me, but here we were rock-solid-still. We'd been told to expect the remaining fuel in the descent stage to slosh back and forth after we touched down, but there simply wasn't enough reserve fuel to do this."[43]

Duke told them of a potential problem. "Tranquillity, Houston. We have an indication that we've frozen up the descent fuel helium heat exchanger – and with some fuel trapped in the line between there and the valves, and the pressure we're looking at is increasing there. Over."

Armstrong replied, "Roger. Understand."

Just after *Eagle* settled down on the lunar surface, a group of engineers tucked away in the bowls of Houston were reaching a crisis point. Readings on their television screens were showing pressure and temperature were rising in one of the descent stage fuel lines. Tom Kelly from the Grumman factory that built the Lunar Module, was sitting in an interface area called SPAN (Spacecraft Analysis) watching his screen with increasing alarm as he saw the residual heat from the descent engine, which had just been shut off, creeping up to a slug of frozen fuel left in the pipe. He was afraid that the fuel would become unstable when heated and the consequence would be an explosion like a hand grenade, which would probably damage the ascent stage.

As the pressure built up in the fuel line, so did the pressure on the engineers, hanging grimly onto their phone handsets, discussing what to do next. What should they do? To abort the mission at this stage was unthinkable, but so was stranding or injuring the astronauts. Could they burp the engine for a moment to try and clear the plug? Just when the tension was becoming unbearable the pressure began to drop before their eyes, rose, and fell away. The engineers relapsed into a state of relieved shock. The mission could continue and the world outside could carry on celebrating. In the Lunar Module Armstrong and Aldrin felt that the worst that could happen would be a fracture in the pipe. As they had finished with the descent engine they did not consider it a serious problem.

During the same morning at Arlington National Cemetery a bunch of flowers appeared on John F. Kennedy's grave. A note with them said: "Mr President, the *Eagle* has landed."

But where had they landed? Nobody was sure.

Armstrong didn't know. "Houston. The guys who said we wouldn't be able to tell precisely where we are, are the winners today. We were a little busy, worrying about program alarms and things like that in part of the descent where we would normally be picking out our landing spot, and aside from a good look at several craters we came over in the final descent, I haven't been able to pick out the things on the horizon as a reference yet."

Duke reassured him, "Roger, Tranquillity. No sweat. We'll figure it out. Over."

But it wasn't that easy – the mapping people were sweating now. Collins in *Columbia* was vainly scanning the lunarscape for signs of the Lunar Module each time he passed over, guided by Houston's latest update from the Mapping Sciences Laboratory in Houston. Using huge lunar maps and data from the spacecraft and tracking stations they narrowed it down to a 8 kilometres radius. Armstrong and Aldrin could not identify anything of significance from their position. It wasn't until they were half way home that their position was pinpointed by a chance remark by Armstrong.

The Madrid and Goldstone tracking stations covered the lunar landing, while in Canberra it was 6:18 am and the Honeysuckle operations crew were just beginning to prepare their equipment for the upcoming day's epoch making events in what was called the SRT, or Site Readiness Test.

Craterland

Scanning the view through the Lunar Module's window on the lunar surface, Armstrong reported his first impressions of the lunarscape. "We'll get the details of what's around here, but it looks like a collection of just about every variety of shapes, angularities, granularities, every variety of rock you could find. The colours vary pretty much depending on how you are looking. There doesn't appear to be too much of a general colour at all, however it looks as though some of the rocks are boulders, of which there are quite a few in the near area, it looks as though they're going to have some interesting colours to them, over."

The flight plan called for a four hour rest period after landing. As everything had gone according to schedule, the Lunar Module was in good shape, and the astronauts weren't admitting to being tired, they were very keen to get out before their rest period.

Armstrong explained, "We wanted to do the EVA (lunar walk) as soon as possible. It would make more sense to go ahead and complete the EVA while we were still awake and not try to put that activity in the middle of a sleep period."

So at the time, Armstrong suggested, "Our recommendation at this point is planning an EVA, with your concurrence, starting about 8 o'clock this evening, Houston time. That is about three hours from now."

Duke: "Stand by."

Armstrong: "Well, we will give you some time to think about it."

Duke: "Tranquillity Base, Houston. We thought about it. We will support it. We're *go* at that time. Over."

Armstrong: "Roger."

The first two hours were very busy for everybody as the two astronauts and Mission Control ploughed through reams of procedures to prepare for the moonwalk.

At Carnarvon the town had no television, so the Australian Broadcasting Corporation had made arrangements with the Overseas Telecommunications Commission to receive the Moon walk television via satellite. A 36 cm monitor was installed in the local theatre, the people at the back of the crowded hall using binoculars and rifle telescopes to see the history making moments.

Honeysuckle Creek had finished all the testing and setting up, begun over five hours earlier at 6 am, and went into the H-30 count, 30 minutes to the signal from the spacecraft appearing on the horizon with moonrise. While a freezing-cold westerly wind dragged sleet showers over the valley, the 26 metre antenna dropped down to the horizon and waited, servos whining. Everybody and everything in the station was ready waiting, waiting for the first signs of a signal from the lunar spacecraft. Tidbinbilla would have a bit longer to wait for the Command Module to appear from behind the Moon on its seventeenth orbit.

At 11:15 am local time, the Moon rose above the gum-tree clad horizon beside Dead Man's Hill until the Honeysuckle Creek receivers promptly locked on to the *Eagle*'s signals. They flooded through the station's equipment, kicking meter pointers up scales, rolling figures around readouts. Anxious eyes rapidly scanned over the panels, watching until they all settled nicely down to normal readings. Everything was working perfectly. Selected data was sent on to Mission Control at Houston in Texas.

The Moon had been rising over that horizon for millions and millions of years – but this time it was different, it was unique – this time men from Earth were there. The astronauts were getting ready to climb out of the Lunar Module – it was 1 hour 41 minutes to Armstrong stepping on the Moon's surface. The atmosphere in the humming equipment rooms at Goldstone and Honeysuckle Creek was charged, charged with a spine tingling tension you could feel – like goose pimples on the arms.

John Saxon, Operations Assistant at Honeysuckle, said, "The checks on the portable life support systems at this point were in a totally different sequence to what we were expecting – every time they changed modes we had to make major reconfigurations on the ground – we were really, really busy trying to keep up with the astronauts doing their own thing. The busiest man without question, was Kevin Gallegos, the man at the front end at the Sub Carrier Demodulation equipment because all these modes affected how he routed the signals through the station and he had to literally second guess what the astronauts were doing, because they were not following the planned sequence. He was calling out all the things he was seeing, we were then directing the telemetry people who were actually processing the support data such as the astronauts heart beats, trying to report to Houston what was happening all the time, telling them what we were doing, and keeping a log of all the events. It was a real team effort."

Kevin Gallegos said, "I broke out in a cold sweat because they were saying things should be happening, the Saxons of the world were saying it had happened, the telemetry bloke was looking around saying 'where is it?' and I looked up and found to my horror that I wasn't patched right. It was one of those things you have gone over that many times in your head – I just looked at it in utter disbelief so I quickly whipped it around, and in the euphoria of the moment everything took off. There's ten seconds of my life there where I can still feel that cold sweat twenty five years later!"

Meanwhile at Houston, hub of the Apollo wheel, nexus to the world's Apollo media, this was the moment of truth. They were all gathered in the theatre, staring at the large black screen as the astronauts struggled into their suits. Their muffled voices could be heard as they prepared for the Moon walk, while the black screen just mocked the journalists' increasing impatience. It took nearly a whole normal working day for the two astronauts to get ready to climb out of the hatch.

At this point *Columbia* went out of sight behind the Moon. Collins desperately wanted to hear what Armstrong was going to say when he stepped on the Moon and he realised he was the only person out of contact with the epoch-making events. All the billions of people around Earth and the two on the other side of the Moon, and he was the only person completely cut off from it all! Complete silence except for the spacecraft noises. *Columbia* did not reappear until Armstrong and Aldrin were raising the flag, so Collins missed hearing Armstrong's momentous step onto the lunar surface.

Armstrong: "Everything is go here. We're just waiting for the cabin pressure to bleed, to below enough pressure to open the hatch."

Houston: "Roger, we're showing a real low static pressure on your cabin. Do you think you can open the hatch at this pressure?"

Armstrong replied, "We're going to try it. ..." At first it resisted their efforts to open it, but when Armstrong pulled back a corner of the hatch and broke the seal – it yielded: "... hatch coming open."

Aldrin crouched inside the Lunar Module and guided Armstrong out onto the porch. After reaching the platform, Armstrong crawled back into the cabin to check he could return safely, before tackling the ladder.

> Aldrin: "OK, you're not quite squared away … roll right a little … now you're even."
>
> Armstrong: "OK, that's OK."
>
> Aldrin: "That's good. You've got plenty of room to your left. It's a little close on the. …"
>
> Armstrong: "How am I doing?"
>
> Aldrin: "You're doing fine. OK. Do you want this bag?"
>
> Armstrong: "Yeah … got it. …"

Aldrin handed Armstrong a bag of rubbish to throw down onto the lunar surface. Armstrong proclaimed, "OK, Houston. I'm on the porch."

That First Step

Sitting at his Director of Flight Operations console with Bob Gilruth and George Low, Chris Kraft was looking at the climax of his life's work. As the moment approached his mind, as always, was looking ahead, looking for possible emergencies and instant solutions and thinking: "Is Armstrong going to have trouble walking? Is the suit manoeuvrability going to be enough? Is that damned TV camera going to work? What is Armstrong going to say? When is he going to tell me how it is to walk in one sixth G?"[37]

The watching world paused, and everyone had those same thoughts – those same questions. Bruce McCandless was the Capcom at Houston.

Armstrong began to carefully work his way down the ladder when Aldrin reminded him to open the MESA (Modular Equipment Stowage Assembly) door to expose the television camera. "Did you get the MESA out?"

Figure 5.6. In Australia on the morning of the first lunar walk, Prime Minister John Gorton (left) visited Honeysuckle Creek to see the action for himself. Station Director Tom Reid explains the operation of the antenna.

Armstrong reached out to yank the D-ring and lanyard. "I'm going to pull it now. Houston, the MESA came down all right."

Houston: "This is Houston. Roger, we copy. We're standing by for your TV."

Armstrong: "Houston, this is Neil. Radio check?"

Houston: "Neil, this is Houston. You're loud and clear. Break. Break. Buzz, this is Houston. Radio check and verify TV circuit breaker in."

Aldrin: "Roger. TV circuit breakers in and read you loud and clear."

Houston: "Roger ... and we're getting a picture on the TV."

Aldrin: "You got a good picture, huh?"

Houston: "There's a great deal of contrast in it, and currently its upside down on our monitor, but we can make out a fair amount of detail."

A contrasty picture full of fuzzy black and white shapes suddenly appeared on the Houston TV monitors and the big screen in the theatre – everyone cheered wildly, then peered at it, fascinated by the knowledge that they were witnessing a supreme moment in history but at first unable to resolve the strange shadows and bright highlights offered by the screen.

On Earth there was a flurry of energetic activity as the television operators tried to decipher the strange fuzzy shapes on their screens. This was the conversation on the inter-station communication channel.

Goldstone: "TV on line, Goldstone TV on line."

Honeysuckle: "Honeysuckle video on line."

Houston TV: "Houston TV, we have both sites five by."

Houston TV: "Goldstone, can you confirm that your reverse switch is in the proper position for the camera being upside down?"

Goldstone: "Stand by, we will go to reverse position. We are in reverse."

Houston TV: "Roger, thank you. All stations, we have just switched video to Honeysuckle."

The first picture used was being sent by the big 64 metre Mars antenna at the Goldstone Tracking Station in California, but they were in trouble. Bill Wood, USB Lead Engineer at Goldstone, explained, "It was a much stronger signal coming down from the Lunar Module than we had expected, it ran into clipping. Since the signal was also inverted – that is white on black instead of black on white, and as the clipping was on the black side, the picture was coming down to us almost completely black, very little white, there was no detail. I saw the network TV here – we were picking up the commercial television out of Los Angeles and when we saw the switch from Goldstone to Honeysuckle there was a pronounced improvement in the video quality. 'Hey, look at the picture from Honeysuckle!' and I thought 'Good Lord there's something wrong with our system – they are getting it much better than we are'."

Ed von Renouard, the television technician at Honeysuckle Creek, added, "When I was sitting there in front of the scan converter, waiting for a pattern on the input monitor, I was hardly aware of the rest of the world. I heard Buzz Aldrin say 'TV circuit breaker in' and at the same moment I saw the sloping strut of the Lunar Module's leg against the Moon's surface. The input monitor receiving the slow scan from the spacecraft showed the scene upside down, but this was expected and planned for, and our signal went to Sydney right way up. A few weeks before the mission someone at NASA discovered that when the MESA hatch with the TV camera attached was opened the camera would be upside down, so a simple switch was installed at the tracking stations to invert the

Figure 5.7. Ed von Renouard at the Honeysuckle Creek Television Console. The slow scan television processing rack used to send the first pictures of Armstrong setting foot on the lunar surface is on the left.

picture. When the image first appeared in front of me it was an indecipherable puzzle of stark blocks of black at the bottom and gray at the top, bisected by a bright diagonal streak. I realised that the sky should be at the top, and on the Moon the sky is black, so I reached out and flicked the switch and all of a sudden it all made sense, and presently Armstrong's leg came down."

It was about six minutes too early for the big Parkes 64 metre dish to pick up the Lunar Module's signal properly, but the 26 metre antenna at Honeysuckle Creek was firmly locked on. Houston video switched to the Honeysuckle Creek signal, though they left the sound from Goldstone, and viewers around the world were able to make out the ghostly looking scene of the black sky and white lunar surface. First of all there was no movement.

Then … on millions of screens around the world Earthlings saw a fuzzy big blob detach itself from the top left corner, and could make out a leg seeking the next rung of the ladder. Above the leg was a whitish suggestion of the rest of Armstrong.

Houston called, "OK, Neil, we can see you coming down the ladder now."

In the shadow of the Lunar Module Armstrong worked his way down the nine steps, carefully placing his feet on each rung. At the bottom rung he found he was still more than a metre above the lunar surface, mainly due to the gentle landing not compressing the spacecraft's legs. He felt he could manage the drop so tried a measured leap aimed at the Lunar Module's footpad, landing comfortably on both feet. To confirm he could get back up with the restrictions of the suit he promptly jumped back onto the bottom rung.

Armstrong: "Okay, I just checked – getting back up to that first step, Buzz. The strut isn't collapsed too far, but it's adequate to get back up.

Houston: "Roger. We copy."

Armstrong: "It takes a pretty good little jump."

After checking camera settings, the moment to consummate the whole Moon landing programme was at hand.

Neil Armstrong spoke those momentous words: "I'm at the foot of the ladder. The LM footpads are only depressed in the surface about one or two inches, although the surface appears to be very, very fine grained, as you get close to it. It's almost like a powder. The ground mass is very fine … I'm going to step off the LM now. …"

Just 12 years after Sputnik shook the world, at 9:56:20 pm Houston Daylight Time, 20 July (12:56:20 pm Australian EST, 21 July 1969) 38-year-old Neil Alden Armstrong from Wapakoneta, Ohio, USA, dropped back onto the footpad and lifted his left foot backwards over the lip to test the lunar soil, making furrows in the dust with the toe of his boot. As it seemed firm enough he moved off the footpad and let go of the *Eagle* to create one of the greatest moments in the existence of mankind, and to consummate the work of the 400,000 people that were behind putting those boots there, and beyond them to all the visionaries back through history that had a dream, however wild and improbable at the time, of man among the stars. From now until the end of time, we can regard ourselves as people *from* the planet Earth.

To Buzz Aldrin, poised on the top of the ladder, it seemed a small eternity before he heard Armstrong say:

"That's one small step for (a) man, one giant leap for mankind."

Armstrong commented later, "I had thought about what to say a little before the flight, mainly because so many people had made such a big point of it. I had

Figure 5.8. While the world held its breath watching Armstrong step onto the lunar surface on their television screens…

Figure 5.9. … Collins was completely cut off from humanity on the other side of the Moon. Although dying to know what Armstrong was going to say, he didn't reappear until Armstrong and Aldrin had raised the flag.

also thought about it a little on the way to the Moon, but not much. It wasn't until after landing that I made up my mind what to say.

"I don't recall any particular emotion or feeling other than a little caution, a desire to be sure it was safe to put my weight on that surface outside *Eagle*'s footpad."

John Saxon at Honeysuckle Creek, said, "I wrote a few things in red ink in the log, like 'Touchdown!!' What I should have written when Armstrong stepped on the Moon's surface was: 'Commander on the surface.' I was so excited I wrote: 'Commander on the Moon!!!!', and the time to the second because we had a big office sweepstake going on the exact time he stepped out."

Tom Reid, Honeysuckle Creek Station Director, said, "There were four contingencies which resulted in Honeysuckle Creek being the station which sent the picture of Neil Armstrong's footstep around the world.

"First of all the original Flight Plan called for the egress to occur when the Goldstone and Parkes 64 metre antennas were in view, so there would be 100 per cent redundancy in 64 metre antennas. Armstrong, however decided to come out early, and the Mission Controllers decided they wouldn't oppose that. Because of that, when they actually did come out Parkes didn't have a view because they had an elevation constraint (the Moon wasn't high enough for the Lunar Module's signals to enter their beam). Unfortunately, at the same time there was a problem at Goldstone, and they were getting poor slow scan TV back. It was upside down as well, and due to the transmitter failure earlier at Tidbinbilla, Honeysuckle Creek was tracking the Lunar Module."

First Walk in Another World

The world watched breathlessly, straining to hear every word from the two astronauts, waiting for the first descriptions of a lunar experience, first descriptions of another world. After gaining confidence by bouncing up and down a few times, Armstrong stepped sideways, let go of the Lunar Module's landing strut, and was standing by himself on the Moon's surface.

Armstrong described what he found. "The surface is fine and powdery. I can kick it up loosely with my toe. It does adhere in fine layers, like powdered charcoal, to the sole and sides of my boots. I only go in a small fraction of an inch, maybe one eighth of an inch, but I can see the footprints of my boots and the treads in the fine, sandy particles."

Houston replied, "Neil, this is Houston. We're copying."

Then after a pause, Armstrong continued, "There seems to be no difficulty in moving around, as we suspected. It's even perhaps easier than the simulations of one sixth g that we performed in the various simulations on the ground. It's absolutely no trouble to walk around."

After Armstrong prepared to deploy the camera equipment, he added, "Looking up at the LM, I'm standing directly in the shadow now looking up at Buzz in the window. I can see everything quite clearly. The light is sufficiently bright, backlighted into the front of the LM, that everything is clearly visible."

After some discussion about the cameras, Armstrong said, "I'll step out and take some of my first pictures here."

Houston responded, "Rog., Neil. We're reading you loud and clear. We see you getting some pictures and the contingency sample."

The CSIRO's 64 metre dish antenna at Parkes came on line at 1:02 pm local time when the Lunar Module on the Moon rose high enough above the horizon for its signal to enter the main beam of the big dish and Sydney Video advised Houston TV:

Sydney Video: "Houston TV, Sydney Video."

Houston TV: "Houston TV, Go ahead."

Sydney Video: "Please be advised I have a very good picture from Parkes, shall I give it to you?"

Houston TV: "Roger."

Sydney Video: "You have it."

Houston TV: "Roger, beautiful picture, thank you."

Houston TV: "We are switching to Parkes at this time."

Network: "Honeysuckle, Network."

Station Director Tom Reid: "Network, Honeysuckle."

Network: "You might pass on to the Parkes people their labour was not in vain, they've given us the best TV yet."

Reid: "Roger, thank you very much, they'll appreciate that, they're monitoring."

Within 15 minutes, verbally guided by Armstrong, Aldrin backed carefully out of the Lunar Module's hatch.

Aldrin: "Okay, Now I want to back up and partially close the hatch … making sure not to lock it on my way out!"

Armstrong chuckled: "A particularly good thought."

Aldrin: "That's our home for the next couple of hours, we want to take good care of it."

Careful not to fall off the porch, Aldrin worked his way down the ladder and jumped onto the Lunar Module's footpad. He also tried to leap back onto the ladder, but missed the bottom rung by about two centimetres, so he tried again.

Armstrong: "There, you've got it."

Aldrin: "That's a good last step!"

Armstrong: "Yeah, about a three-footer."

Aldrin: "Beautiful view!"

"Isn't that something! Magnificent sight out here," Armstrong greeted Aldrin as he joined him on the lunar surface.

Figure 5.10. "Be sure not to lock it (the hatch) on the way out!" jested Aldrin as he climbed down to join Armstrong on the lunar surface.

"Magnificent desolation," returned Aldrin, inspired by Armstrong's comment.

Aldrin later explained, "Stepping out of the Lunar Module's shadow was a shock. One moment I was in total darkness, the next in the Sun's hot floodlight. I stuck my hand out past the shadow's edge into the Sun and it was like punching through a barrier into another dimension."[43]

The two astronauts found the spacesuits very comfortable with little interference to their mobility, except when bending down to pick up objects from the lunar surface. The suits were designed to cope with the extreme conditions expected in the lunar environment, isolating the astronauts from the vacuum outside and the wildly fluctuating temperatures. The temperature of the ground they were walking on could vary from 110°C in the sunlight to -170°C in the shade.

Armstrong said he was not aware of any temperature changes inside the suit while he touched objects or walked about. "From inside the cockpit the Moon looked warm and inviting. The sky was black but it looked like daylight out on the surface, and the surface looked tan. There is a very peculiar lighting effect on the lunar surface that seems to make the colours change. If you look down-Sun, down along your own shadow, or into the Sun, the Moon is tan. If you look cross-Sun it is darker, and if you look straight down at the surface, particularly in the shadows, it looks very, very dark. When you pick up material in your hands it is also dark, gray or black.

"The material is of a generally fine texture, almost like flour, but some coarser particles are like sand. Then there are, of course, scattered rocks and rock chips of all sizes."

Aldrin impressions were: "I felt buoyant and was full of goose-pimples. I quickly discovered that I felt balanced – comfortably upright – only when I was tilted slightly forward. I also felt a bit disorientated: on the Earth when one looks at the horizon, it appears flat; on the Moon, so much smaller than the Earth and quite without high terrain, the horizon in all directions visibly curved away from us."[43]

After taking the first photographs, Armstrong moved into the sunlight and began collecting the first soil samples. The equipment on his back had weighed 38 kilograms on Earth, but on the Moon it was only 6.3 kilograms.

Armstrong said, "After landing we felt very comfortable in the lunar gravity. It was, in fact, in our view preferable both to weightlessness and to the Earth's gravity."

Aldrin added, "One-sixth gravity was agreeable, less lonesome than weightlessness, I had a distinct feeling of being somewhere."

Armstrong felt they had landed in a timeless place, with no changes to mark time passing as we know it. Although the astronauts were locked into the time in Texas, here at Tranquillity the scene would have been just the same a thousand years ago, and probably the same a thousand years in

Figure 5.11. The eye of the television camera on the Lunar Module witnessed a moment of pride as Neil Armstrong (left) held the staff, while Buzz Aldrin set America's national flag flying on the Moon's surface to consummate the primary goal of the Apollo Program.

the future. With no atmosphere, they found that everything they could see was starkly clear; features on the horizon were as sharp and clear as the rocks at their feet.

Aldrin said, "I looked high above the dome of the Lunar Module. Earth hung in the black sky, a disk cut in half by the day night terminator. It was mostly blue, with swirling white clouds, and I could make out a brown land mass, North Africa and the Middle East. Glancing down at my boots, I realised that the soil Neil and I had stomped through had been there longer than any of those brown continents."[43]

Armstrong and Aldrin set up the Stars and Stripes flag, finding it difficult to lunch the pole into the lunar soil, or lurain as some call it. They only managed to sink it into the soil about 20 centimetres, then lean it so the weight of the flag didn't pull it over.

In Washington President Richard Nixon was watching the moonwalk in the White House, and remembered Frank Borman from Apollo 8, Bob Haldeman, and he were standing around the TV set in his private office when they watched Neil Armstrong step onto the Moon. Then he went into the Oval Office next door where the media TV cameras had been set up for his split screen phone call to the Moon. Armstrong's voice came through loud and clear.

The president said, "Because of what you have done the heavens have become a part of man's world. And as you talk to us from the Sea of Tranquillity, it inspires us to redouble our efforts to bring peace and Tranquillity to Earth. For one priceless moment, in the whole history of man, all the people on this Earth are truly one. One in their pride in what you have done. One in our prayers that you will return safely to Earth."

Figure 5.12. Buzz Aldrin looked thoughtfully at the Stars and Stripes: "I was dreading the possibility of the American flag collapsing into the dust in front of the television camera." The flag was not a territorial claim, said NASA, it was to identify the nation that achieved the landing.

Away up beyond the sky, the men on the Moon paused and listened to their President. Armstrong responded with, "Thank you Mr. President. Its a great honour and privilege for us to be here, representing not only the United States but men of peace of all nations, and with interest and a curiosity and a vision for the future. It's an honour for us to be able to participate here today."

President Nixon continued, "Thank you very much, and I look forward, all of us look forward to seeing you on the *Hornet* on Thursday."

Aldrin replied, "I Look forward to that very much, sir."

They saluted the camera sitting there on its tripod in that desolate, empty wasteland, and turned to the job of collecting the samples of Moon rocks.

While most of the world followed the mission on the media, Russia and China had little to say, in fact as the Lunar Module was dropping down to the Moon's surface Moscow television was broadcasting a film on the life of a long dead Polish singer, and Moscow Radio was reviewing the week's sport. Moscow television referred to the landing in its final news broadcast of the day.

Armstrong threw a rock with the comment, "You can really throw things a long way out here." As he darted about collecting the rock samples, his pulse rate went up to 140, peaking at 160 when he hauled the samples up into the Lunar Module with a special block and tackle he called the "Brooklyn clothes line."

An important part of the Apollo missions was to leave a scientific package on the Moon's surface for the tracking stations on Earth to monitor the conditions around the landing site after the astronauts left. Apollo 11's package was called EASEP (Early Apollo Scientific Experiments Package), while the remaining lunar landing missions left a more elaborate package called ALSEP (Apollo Lunar Scientific Experiments Package). The instruments measured particles from the sun, the Moon's seismic activity, and a laser beam reflector for accurately measuring the distance between the Earth and Moon. Carnarvon's 9-metre dish was scheduled to track EASEP, at times it was the prime support.

The superb colour photographs taken by Armstrong using a newly introduced Hasselblad 500 EL Data Camera, while occasionally equalled, were never surpassed by subsequent missions. The people at Hasselblad held their breath during Apollo 11, because the astronauts only used one camera, and if it failed there would be no pictures! Wally Schirra took the first Hasselblad camera into space in his Mercury MA-8 flight, a slightly modified off-the-shelf model 500 C purchased from a local store. NASA then approached Hasselblad to develop a camera for use in the space environment – they wanted at least 5000 working cycles in the Earth's atmosphere, in pure oxygen and in a vacuum, using a 70 mm film magazine made for thin film which gave 160 exposures to a roll. A glass plate in the exposure frame engraved with 25 crosses allowed precision calculations to be made. The Apollo 11 camera had the mirror and focussing plate removed, and a specially designed Zeiss Biogon f5.6 60 mm focal length lens painted with an aluminium coating to control heat absorption and radiation. Ten of these cameras are now lying on the lunar surface. Some 33,000 pictures were taken during the Mercury, Gemini and Apollo missions by 52 Hasselblad cameras, covering a total distance in space of 65.3 million kilometres.[56]

Who is the astronaut in all those still pictures from Apollo 11, now among the most famous images of the Twentieth Century? They are all of Aldrin. An intensive study has revealed that there is only one frame showing a back view of Armstrong by the Lunar Module, and some indistinct frames showing parts of him mixed with the Lunar Module structure.

On one of the Lunar Module's legs is a plaque signed by President Nixon and the three astronauts to commemorate the landing. It is ironical that President Nixon's name goes on the plaque on the Moon, when it was Kennedy and Johnson who did so much to bring the event about.

Back to the Lunar Module

At 2:11 am the hatch was closed to complete a 2 hour 47 minute 14 second moonwalk. As the two astronauts struggled with the rocks and suits in the cramped Lunar Module cabin there was a lengthy pause in communications.

Dr Ross Taylor, Principal Lunar Sample Investigator from the Australian National University in Canberra, explains, "Everyone knows if you take iron filings – this is a high school chemistry experiment – and put them in an oxygen atmosphere they will burst into flame. This had been raised as a possible hazard to the astronauts. The samples were supposed to have been sealed up in a box, but of course dust had got everywhere, although we didn't know that at the time. I was watching the big screen at Houston and finally they got back in. We sat there in silence and waited for some minutes, it seemed like a very long time, and nothing came through until finally mission control called up Tranquillity Base, and the first time there was no answer they called a couple more times – you could hear from the controller's voice there was a certain amount of tension at the time – I thought, I hope these lunar samples had not ignited in the pure oxygen atmosphere when they repressurised the Lunar Module, so I sat through rather a bad five minutes until finally they came on the air to say they had been getting out of their suits. A couple of my colleagues had looked at one another, also wondering if this had happened."

Armstrong remembers, "When we got back in the Lunar Module, closed our hatch, repressurised, and took our helmets off, there was a decided odour in the cockpit. To me it seemed like the odour of wet ashes in a fireplace. I can't be certain that it came from the lunar material, although that would be my guess."

Aldrin adds, "There was a faint metallic gunpowder odour to the moonrock. It wasn't at all overpowering, but it was noticeable. It did not tempt us to stick our nose right up against it and inhale."[43]

Ross Taylor says, "To my knowledge the lunar soil would not have a smell, and I have handled plenty of it. I suspect there is nothing in it to give rise to a smell as it is totally dry and there is no organic material in it."

The smell sensed by Armstrong and Aldrin was probably the lunar soil first coming into contact with the oxygen atmosphere of the Lunar Module.

Aldrin continues, "Our first chore was to pressurise the Lunar Module cabin and to begin stowing the rock boxes, film magazines, and anything else we wouldn't need until we were connected again with the *Columbia*. Following that, we removed our boots and the big back packs, opened the Lunar Module hatch and threw these items onto the lunar surface. The exact moment we tossed everything out was measured back on Earth – the seismometer we had put out was even more sensitive than we had expected."[43]

With no airlock to isolate the cabin, they breathed with oxygen fed into their helmets while they threw the unwanted material out.

They expected the lunar dust particles to float around inside the Lunar Module, but were surprised to find that they never did, generally staying where they lodged, probably due to the fact they were so dry they were attracted to anything with static electricity. This meant they were able to remove their helmets without the worry of the dust getting in their eyes and noses.

At 5:25 am the tired astronauts finally put blinds over the windows and curled up to rest. They both decided to sleep with their helmets and gloves on, hoping there would be

Figure 5.13 (inset left) Buzz Aldrin in the Lunar Module after the moonwalk...
Figure 5.14 (inset right) ... while Armstrong looked tired but very pleased with their successful efforts.
Figure 5.15 (background) The lunar far side, from Apollo II near Astronomical Union Crater No 312.

less noise, they would be warmer, there was less chance of breathing lunar dust, and they would not have to find somewhere to stow them.

The later missions supplied hammocks for sleeping, but not for Apollo 11, so Aldrin lay on the floor with his feet up against the side, as the cabin wasn't wide enough to stretch out. Armstrong sat on the cover of the ascent engine, leaned against the rear of the cabin, and suspended his legs through a loop of waist tether he had rigged up from a handhold. After he settled down Armstrong found there was an annoying pump gurgling somewhere near his head, and he could not avoid seeing the Earth glaring at him like a big blue and white eyeball through the Alignment Optical Telescope, so he had to get up and block the Earth light off.

Both agreed they did not sleep very well. Apart from the emotional high from excitement of the day they became progressively colder, though they tried turning the suit water temperature up to maximum, then disconnecting the water flowing through their suits. Aldrin finally adjusted the temperature of the air flow through the suit and they felt better. The cabin temperature was steady at 16°C. Dr Kenneth Biers at Houston said the data he received from Armstrong – they were not monitoring Aldrin – indicated that he may have slept fitfully and dozed, but stirred around quite a bit.

How would you sleep after the epoch-making events of the day and wondering about tomorrow? Will the motor work? It had never been fired on the Moon before. What would happen if it didn't fire, or misfired and dumped them back on the surface? What would they do? Their survival depended solely on the tanks of oxygen they had brought with them. If they didn't sleep and were over-tired tomorrow, it would be all the easier to make a mistake. If they were ever exposed to the vacuum outside their spacesuits or spacecraft, their blood would boil in ten seconds.

It is no wonder they didn't sleep very well.

Lunar Launch

At 12:13 pm on 21 July the Lunar Module astronauts were woken up by Ron Evans, the Houston Capcom with, "How is the resting standing up there? Did you get a chance to curl up on the engine can?"

Aldrin responded, "Roger, Neil has rigged himself a really good hammock with a waist tether, and he's been lying on the ascent engine cover, and I curled up on the floor."

As the lunar lift-off time drew near Collins became nervous. His job was relatively simple if the Lunar Module performed to plan, but what if something went wrong? What if the engine didn't fire or they didn't make the proper orbit. He had no means of going down to rescue them, he would just have to leave them behind and come home alone. They only had enough oxygen for a day at the most.

Armstrong recalls, "I thought quite a bit about that single ascent engine and how much depended on upon it. When the moment came it was a picture of perfection."

After 21 hours on the lunar surface, the two lunar explorers prepared their ship for lift off. During the detailed prelaunch checks they suddenly came across a switch with the toggle lever broken off – they couldn't operate it! Aldrin must have broken it off while he was getting into his spacesuit for the lunar walk, as it was on his side of the cockpit. As this switch was the engine arming switch that sent the electrical power to the engine for launching, it meant they were marooned on the Moon unless they could find another way of operating the switch's function. Houston was advised and they broke a similar switch off to seek a fix. Someone realised they had their Fisher space pens at hand so told them to retract the point and use the hollow end. To everyone's relief the pen just fitted into the hole and operated the metal switch, and the launch procedures continued until at last Ron Evans in Houston passed a message up, "Our guidance recommendation is PNGCS, and you're cleared for take off."

Aldrin replied, "Roger, understand. We're number one on the runway."

Right on time at 12:54 pm 21 July the rocket engine that had to fire, fired.

Aldrin relayed the lift-off. "OK, master arm on. Nine … eight … seven … six … five … abort stage – engine arm ascent … proceed. That was beautiful … 26, 36 feet per second up. Be advised of the pitch over. Very smooth … very quiet ride."

Pushed by the 1,587 kilogram thrust rocket for seven minutes, the tiny spacecraft shot up into the black lunar sky, picking up speed from 48 kilometres per hour after 10 seconds to 2,897 kilometres per hour. As the rocket's exhaust gases shredded the gold foil insulation and sprayed the pieces around the landing place, Aldrin looked out of the window long enough to see their flag topple over in the blast from the rocket motor.

Only Madrid was tracking the Lunar Module as it roared into the black lunar sky.

Left behind was an estimated $1 million rubbish dump, the first outside the Earth. Apart from the scorched launching frame and scientific instruments, there were the Stars and Stripes lying in the dust surrounded with empty food packages, cameras, backpacks, a silicon disk with messages from leaders of 73 nations, a gold olive branch, memorials to those Russians and Americans who had died for space exploration, and gear no longer of any use

Aldrin recounts, "Seconds after lift-off, the Lunar Module pitched forward about 45 degrees, and though we had anticipated it would be abrupt and maybe even a frightening manoeuvre, the straps and springs securing us in the Lunar Module cushioned the tilt so much and the acceleration was so great it was barely noticeable."[43] Both astronauts were busy with their respective tasks, Aldrin working on the computer, and Armstrong keeping track of the flight and navigation.

> Evans: "*Eagle*, Houston. You're go at three minutes, everything's looking good."
> Armstrong: "Roger."
> Armstrong: "We're going right down US one."

With the ascent stage of the Lunar Module on its way, the last event that had not been performed before was safely behind, the rest of the mission had been done before by Apollo 10. The two astronauts now began to feel confident that Apollo 11 was really going to make it. There had been no real surprises on the Moon's surface after all.

Eagle Returns to *Columbia*

Up in the Command Module, Collins was preparing to meet his companions with a book of 18 different procedures to rendezvous slung around his neck.

While the Command Module kept a steady course 97 kilometres above the lunar surface, the Lunar Module climbed into a 75.6 kilometres orbit and soon Collins had a

At Tranquility base

Figure 5.16. "For ROCKS?!!..."

radar lock on it, showing it to be 402 kilometres behind. As the Earth waited for the two spacecraft to emerge from behind the Moon, it wasn't long before Collins could see a tiny blinking light in the darkness, then as they passed over the landing site, the Lunar Module was only 24 kilometres below, and 80 kilometres behind. As they entered into sunlight on the back side, Collins saw the blinking light slowly resolve into the Lunar Module skimming over the crater scarred surface below, but looking quite different now without the descent stage and its dangling legs. Armstrong took up a position 15 metres from *Columbia* and kept station. The rendezvous was over and for the first time the astronauts began to feel they were going to bring this amazing stunt off. As they came around the rim of the Moon Houston was agog to know how things were going, but not wanting to interfere with the docking process: "*Eagle* and *Columbia*, Houston standing by."

"Roger, we're station keeping," Armstrong's pithy response told Houston everything. All three astronauts steeled themselves for this critical moment – docking the two spacecraft together again. The success of the mission; their return home; their lives, relied on switches, relays, mechanical latches, and valves all working faultlessly, complementing their own skills. The Lunar Module's docking probe gently entered the cone and with a satisfyingly loud thud the twelve latches slammed home to lock *Eagle* and *Columbia* together again.

Then, just as they began to feel they were safely together again, the spacecraft suddenly began jerking around, both spacecraft thrusters firing in anger. The astronauts all jumped, thinking they might be in trouble, but it was the LM and CSM automatic attitude systems competing with each other until the Lunar Module's automatic pilot was turned off and the spacecraft quietened down to wait for the astronauts' next instructions.

> Armstrong: "OK, we're all yours, *Columbia*."
> Collins: "OK … I'm pumping up cabin pressures … that was a funny one. You know, I didn't feel it strike and then I thought things were pretty steady. I went to retract there, and that's when all hell broke loose. For you guys, did it appear to you to be that you were jerking around quite a bit during the retract cycle?"
> Armstrong: "Yeah. It seemed to happen at the time I put the plus thrust to it, and apparently it wasn't centred because somehow or other I got off in attitude and then the attitude 'HOLD' system started firing."
> Collins: "Yeah, I was sure busy for a couple of seconds."

They were back together again at 4:35 pm, just three minutes behind the time specified by the Flight Plan. Buzz was first through the hatch, with a triumphant grin on his face. Collins gleefully shook his hand, then turned to the tunnel to welcome Armstrong, and an excited reunion took place, before they dragged the lunar rock bags into the Command Module, and prepared for dumping the Lunar Module.

Going behind the Moon for the twenty-ninth time, Collins threw the right switches, and with a slight bang the Lunar Module backed off, watched sadly by Armstrong and Aldrin. Collins, though, was very pleased to see it steadily disappearing into the distance, taking all its complications with it. The *Eagle* would continue to circle the Moon until it finally joined the other spacecraft corpses on the lunar surface.

Coming Home

An orbit later, they carefully lined up the horizon and checked they were in the right attitude before firing the SPS motor on time at 11:56 pm. to set them on a safe course for home.

"Just about midnight in Houston town," commented Armstrong nostalgically.

Honeysuckle Creek's antenna was fastened firmly on the edge of the Moon, waiting for the first signs of a signal. In the spacecraft the astronauts saw the Earth rise above the Moon's horizon for the last time and the voice of Charlie Duke in Houston filled their earphones:

Duke: "Hello. Apollo 11, Houston. How did it go?"

Collins: "Tell them to open up the LRL (Lunar Receiving Laboratory) doors, Charlie."

Duke: "Roger. We got you coming home. It's well stocked."

As they left the Moon, the three astronauts looked back at the huge gray and tan orb suspended in front of them – it was an awesome moment to realise where they were and what they had just done. They tried to use the remaining film to take as many pictures as possible of the moment. Collins, however, felt that he never wanted to return. At 8,000 kilometres from the Moon, the three weary space travellers were able to catch up on their sleep, turning in at about 5:30 am.

After about eight hours' rest, they were left to wake up on their own. They passed through the gravity hump between the Moon and Earth eating their breakfast, 322,021 kilometres from Earth, and 62,553 kilometres from the Moon. The spacecraft now began picking up speed as the Earth's gravity strengthened.

There had been a major effort to try and locate exactly where Apollo 11 had landed in the Sea of Tranquillity, and they were still trying to pinpoint the position when Armstrong dropped a casual remark during a debriefing as they were returning to Earth, "I took a stroll back to a crater behind us that was maybe seventy or eighty feet in diameter and fifteen or twenty feet deep. And took some pictures of it. It had rocks in the bottom."

That description was all the geologists needed – they immediately knew the landing spot from their maps, confirmed by pictures from the 16 mm sequence camera of the landing: 0° 41′ 15″ North latitude, 23° 25′ 45″ East longitude. If only Armstrong had mentioned that crater before!

At 4:56 pm on 23 July the crew celebrated the half way point – 187,000 kilometres to go. During the last evening they sent their final television session, rather a philosophical one. Part of Aldrin's talk said, "We have come to the conclusion that this has been far more than three men on a voyage to the Moon. More still than the efforts of one nation. We feel that this stands as a symbol of the insatiable curiosity of all mankind to explore the unknown."

Armstrong wound his session up with, "… to the agency and industry teams that built our spacecraft – the Saturn, the Columbia, the Eagle and the little EMU, the spacesuit and backpack that was our small spacecraft out on the lunar surface. We would like to give a special thanks to all those Americans who built those spacecraft, who did the construction, design, the tests and put their – their hearts and all their abilities into those craft. To those people tonight, we give a special thankyou, and to all those people that are listening and watching tonight, God bless you. Good night from Apollo 11."

Re-Entry

At 7:47 am on the 24th, the astronauts woke up for their last day in space and prepared for splashdown. They had to separate from the Service Module before they came scorching into the 64 kilometres wide corridor at nearly 40,000 kilometres per hour. The entry

To those of you out there on the Network who made all of the electrons go to the right places, at the right time – and not only during Apollo 11 – I would like to say thank you.

Neil Armstrong
at the Goddard Space Flight Center in Maryland, USA
to the tracking network personnel
18 March 1972

Mission Planning

There are a lot of forgotten people and teams in the 400,000 members of the Apollo Project. Among the leading players were the mission planners, who based their mission timelines on the results of the astronauts' simulations. Considering the complexity and length of the missions, and it had never been done before, the accuracy of their work is astonishing. Splashdown at the end of the mission occurred only 42 seconds behind the planned time!

Rod Rose, chairman of the Mission Planning and Analysis Division of Flight Operations, with Morris Jenkins, from Powered Flight Analysis, offered these comments on the planning of the mission from their view: "There was a big build up of effort by many people in several organisations to obtain a very accurate celestial model. Then we worked out the bugs across the window of opportunity and honed the Earth/Moon relationship. Next, we had very good real-time tracking and navigation, and we designed the mission trajectory profile with allowances for tweaking burns – if needed. We would then make a small RCS correction early, never allowing the mid-course delta-V to build up. The same philosphy was applied on the way back from the Moon."

The Ground Elapsed Time, or GET, is set from the moment of lift off, and allows the whole mission to be planned ahead of the launch, so once the spacecraft leaves the ground, this time can then be related to any time zone on Earth. The list on page 252 shows how little the Apollo 11 mission deviated from the Flight Plan, modified for the decision to take the lunar walk early. This list was published in the NASA Manned Spacecraft Center *ROUNDUP* newspaper, Volume 8, No. 21, dated 8 August 1969, page 3.

The only part of the 110.6 metre tall vehicle to return to Earth, the Command Module, was retired for display to the public in the National Air and Space Museum, Washington, DC.

In 1969 a new titanium bearing mineral was found among the Apollo 11 samples. It was called *Armalcolite* after the astronauts – *Arm*strong, *Al*drin and *Col*lins – but rather lost its uniqueness when some was found on the Earth during the 1970s.

Apollo 11 Moonrock Memorised in Washington Cathedral

Embedded in a specially built window in Washington Cathedral at Mount Saint Alban is piece No. 230, a 0.25 oz (7.18 gram) bit of basalt estimated to be 3.5 billion years old, brought back from the Moon by Apollo 11. Normally cathedral stained-glass windows are divided into lancets, each a separate entity divided by stonework. One of the few windows without human figures, this 19 foot (5.8 metre) high space window uses the entire area, carrying the viewer's eye across the space. The deep colours used in the window reflect the colours in the photographs brought back from space. Orange, red, and

Apollo 11 Event Score Box

Event	GET Planned	GET Actual
Liftoff	00:00:00.0	00:00:00.6
S-IC stage cutoff	00:02:40.4	00:02:41.7
S-II engine ignition	00:02:41.8	00:02:43.0
S-II engine cutoff	00:09:11.0	00:09:08.3
S-IVB engine ignition	00:09:15.0	00:09:12.2
S-IVB engine cutoff	00:11:39.0	00:11:39.2
Insertion 99.4 × 102.6 nmi Earth orbit	00:11:48.8	00:11.49.3
Translunar injection	02:50:13.4	02:50:13.0
S-IVB CSM separation	03:15:03	03:17:04.6
Midcourt correction #1	11:30:00	unnecessary
Midcourt correction #2	26:44:58	26:44:58.8
Midcourt correction #3	53:55:00	unnecessary
Midcourt correction #4	70:55:00	unnecessary
Lunar orbit insertion	75:49:00	75:49:50.5
LM CSM separation	100:39:50	100:39:51
Descent orbit insertion	101:36:14	101:36:15
Powered descent initiation	102:33:04	102:33:05
Lunar Landing	102:45:05	102:45:39
Lunar crew egress	107:59:00	109:07:36
Lunar crew ingress	111:47:00	111:39:00
unar liftoff	124:21:00	124:22:00
Terminal phase finalization	127:39:39	127:43:08
Second docking	128:00:00	128:03:00
Lunar module jettison	130:30:00	130:09:00
Transearth injection	135:23:41	135:23:42
Midcourse correction #5	150:29:54	150:29:56
Midcourse correction #6	172:00:00	unnecessary
LMidcourse correction #7	192:06:00	unnecessary
Entery interface	195:03:07	195:03:07
LMidcourse correction #7	192:06:00	unnecessary
Begin blackout	195:03:25	195:03:24
End blackout	195:07:00	195:06:59
Main shoot deployment	195:12:56	195:12:57
Splashdown	195:17:53	195:18:35

white spheres swirling in a deep blue and green void are surrounded by tiny particles of clear glass which suggest twinkling stars. A thin white line suggests the trajectory of a spaceship across the heavens. Beneath the window is written, "Is not God in the heights of heaven?" from the Book of Job.

Rodney Winfield, the artist from St Louis, explains, "The idea is that the artist only holds creation, allowing space to flow through him ... the more the channel is open, the greater the flow. When the work sings or communicates, one has been sung through, allowed creation to take its natural form. Ultimately, the creative act become an act of prayer, of union and communion."

The Space Window is the result of four years work and planning by two past NASA Administrators, Dr Thomas Paine and Dr James Fletcher, in association with the Very Reverend Francis Sayre, then Dean of the Cathedral. White House officials had said if they gave it to the Washington Cathedral they would have to give some to every church in the US, but Dr Fletcher said that the cathedral isn't merely a church – it is a national shrine visited by hundreds of thousands of tourists. The moonrock was finally approved by President Nixon in a letter to Dr Paine, who donated the window to the cathedral.

A solemn ceremony was held on 21 July 1974. Watched by his fellow space travellers Buzz Aldrin and Michael Collins, Neil Armstrong presented the piece of the Moon to Dean Francis Sayre. As he handed it across to the dean in the high altar, Armstrong said, "Very Reverend sir, on behalf of the President and the people of the United States we present unto you this fragment of creation from beyond the Earth to be embedded in the fabric of this house of prayer for all people."

Figure 5.22. The plaque left behind was fastened between the third and fourth rungs of the Lunar Module's ladder.

"I hope that some wayward stranger in the third millennia may read it and say 'This is where it all began.' It can be the beginning of a new era when man begins to understand his Universe, and man begins to truly understand himself." *Neil Armstrong*

6 Succession

Spaceflights are not miracles, but are directly related to technological engineering on the ground.

Astronaut James Lovell

After the worldwide euphoria of Apollo 11's success, the Apollo teams all settled down to a regular routine of a mission roughly every two months. However, with Apollo 11 out of the way, there were signs that the end of manned flight as we had known it was in sight, even to those in the lowest ranks. The media began looking into the future – but they couldn't see much there, except ways to cut the horrendous costs with proposals such as the shuttle re-useable spacecraft. Tentative plans and some funding were laid down for a manned mission to Mars, but that was to soon lose momentum.[45]

Alan Blake, Tidbinbilla Transmitter Engineer remarked, "On the first Apollo mission there were people coming from everywhere to sit in the briefing room to watch the two television sets we had in there, and eventually these indistinct, blurred pictures with two white blobs which were supposed to be the astronauts appeared and the people sat there with the eyes glued to the screen – you couldn't get them away. Later on, there were perfectly clear pictures, and when you walked past the briefing room there wasn't a soul in there."

Dr Homer Newell, an Associate Administrator of NASA, said, "After a mere 15 years of space activity, people have become blasé about the subject, and even the most difficult of

◄ **Figure 6.1.** Long, dark shadows dramatise the scene at Hadley Rille in the Apennine Mountains. From its lofty height of 3.9 km, Mount Hadley Delta frowns down on Jim Irwin as he prepared the Lunar Rover for the first vehicular excursion on the moon. Mission Commander Dave Scott was moved to write: "Ours was the first expedition to land among lunar mountains. Never quickened by life, never assailed by wind and rain, they loom still and serene, a tableau of forever. Their majesty overwhelms me."

accomplishments seem to be taken for granted, and extensive benefits in the fields of weather, communications, environment and resources tend to fade mentally into the background of the commonplace."[13]

Author, " I remember that for the tracking stations, though, it was an extremely busy period; there was little time to wonder what the future held – anyway that was somebody else's worry. We had to get ready for Apollo 12."

Figure 6.2. "Waal, we beat 'em again, Al!" After a successful session in the Lunar Module simulator, veteran astronaut Pete Conrad (left) relaxed and savoured the moment with Alan Bean, his friend and Lunar Module pilot for Apollo 12. Conrad knew the Lunar Module like his own backyard as he had practically designed the cabin layout. Despite his distracting humming and whistling while he worked, he was one of the smartest astronauts in the simulator, able to handle the toughest problems the simulation team could find to throw at him. He died in 1999 in a motor cycle accident.

Apollo 12

Originally the Lunar Module pilot for Apollo 12 was listed to be Clifton Williams. Alan Bean was scheduled for the Apollo Applications Program, to follow Apollo. At the time Conrad and his crew were training for the first lunar landing as back up crew, but the game of musical chairs was still playing, and the music stopped again on 5 October 1967. Williams was flying home to see his dying father when his T-38 jet went into an uncontrollable roll and crashed, too low for his parachute to save him. Bean couldn't believe his ears when he heard his old mate Conrad asking him to join his crew as Lunar Module pilot.

The mission insignia showed an American clipper ship for a navy crew and to anticipate the spacecraft providing a means of travelling between the planets the way ships opened the seas to commerce. The four stars represented the crew and Clifton Williams. Pete Conrad, Commander of the mission, recalls, "A lot of people thought we named the spacecraft after naval vessels like the USS *Intrepid* – which we did not. There was a lot of controversy over the names because the military was not too popular in those days in the United States and some people accused us of using military names for our spacecraft when in fact they did not have the proper knowledge. North American Rockwell built the Command Module and we had people out there submit names for the spacecraft with twenty-five words why the name. We had them do the same thing at Grumman Aircraft for the Lunar Module. I wanted to let the people that built them name them. *Yankee Clipper* was named after the US clipper ship, one of the first US ventures around the world in the maritime world. The guy at Grumman named the Lunar Module *Intrepid* based on the Webster's Dictionary definition of the word. We then picked the final names out of the lists."

Chris Kraft, Director of Flight Operations, says, "Launch has always been an uneasy time for me, and I always looked forward to successful separation from the booster.

Mission Data: Apollo 12

Date:	14–24 November 1969
Craft names:	*YANKEE CLIPPER* (CSM) *INTREPID* (LM)
Craft numbers:	AS-507/CSM-108/LM-6
Personnel:	Charles Conrad Jr Richard Gordon Jr Alan Bean
Duration:	10 days 4 hours 36 minutes 25 seconds
Features:	45 Lunar orbits in 3 days 17 hours 2 minutes 31 hours 31 minutes on Lunar surface
EVAs:	2, totalling 7 hours 46 minutes
Lunar samples:	34 kg
CSM weight:	28,830 kg
LM weight:	15,224 kg
Landing area:	3.12° S, 23. 39° W in the Oceanus Procellarum or Ocean of Storms
LM impact:	20 November 1969 at 3.94° S, 21.2° W
Distance travelled:	1,533,704 km

When one adds to this an apprehension caused by bad weather over the Cape, I become even more concerned. It turned out that all of the elements were present for Apollo 12."

President and Mrs Nixon were among the large crowd waiting to see the launch, the only time an American President in office witnessed an Apollo launch. As if to prepare this crew of navy aviators for the Ocean of Storms, the launch area was blanketed by rain when Apollo 12 launched into the overcast stratocumulus cloud with a ceiling of only 640 metres above the ground. Rising from Pad 39A at 11.22 am EST in defiance of Mission Rule 1-404, which said no vehicle shall be launched in a thunderstorm, the huge Saturn V vanished into the murk. Observers then saw two bright blue streaks of lightning – right where the rocket had been. Pete Conrad showed why top test pilots are different from the rest of us when 36 seconds after liftoff, at a height of 1,859 metres, they were hit by lightning. At 52 seconds they were hit again. The control panel indicators went haywire and the attitude ball began pitching. If the vehicle really was beginning to fly erratically there were only seconds before it would break up and explode.

The abort handle was waiting at Conrad's elbow, but he calmly announced to the ground controllers, "Okay, we just lost the platform, gang. I don't know what happened here. We had everything in the world drop out ... fuel cell, lights, and AC Bus overload, one and two, main bus A and B out. Where are we going?"

With the master alarm ringing in his ears, Alan Bean thought he knew all the spacecraft's electrical faults, but looking along the panel of glowing warning lights he couldn't recognise any of them – he had never seen so many lights before.

Conrad remembers, "I had a pretty good idea what had happened. I had the only window at the time – the booster protector covered the other windows – and I saw a little glow outside and a crackle in the headphones and, of course, the master caution and warning alarms came on immediately and

Figure 6.3. "We had everything in the world drop out... where are we going?" called Pete Conrad as lightning struck their Saturn V, 36 seconds after liftoff, and their AC power failed. Cooling water can still be seen tumbling down the just vacated launch tower.

I glanced up at the panel and in all the simulations they had ever done they had never figured out how to light all eleven electrical warning lights at once – by Golly, they were all lit, so I knew right away that this was for real.

"Our high bit rate telemetry had fallen off the line so on the ground they weren't reading us very well on what was happening, so they got us to switch to the backup telemetry system. The ground then got a look at us and they could see that a bunch of things had fallen off the line, but there weren't any shorts or anything bad on the systems so we elected to do nothing until we got through staging. When we got through staging then we went about putting things back on line."

Down among the consoles in the Mission Control Center the steady flow of glowing figures from the spacecraft filing past on the screens were suddenly replaced by a meaningless jumble of characters. All the telemetry signals had dropped out!

John Aaron was the EECOM, the Flight Controller in charge of the Command and Service Module electrical system, and he recalled, "You must remember we did not have a live television view of the launch. I was just looking at control screens which only had data and curves on them. The first thing I realised was we had a major electrical anomaly. But I did recognise a pattern. When we trained for this condition with our simulators it would always read zeros. It so happened that a year before I was monitoring an entry sequence test from the Kennedy Space Center, and the technicians inadvertently got the whole spacecraft being powered by only one battery. I remembered the random pattern that generated on the telemetry system, and for some reason just filed it off to the back of my mind. I did go in the office the next day to reconstruct what happened and found this obscure SCE (Signal Condition Equipment) switch. Few people knew it was there, or what it was for. It was lucky I was the EECOM monitoring the test that night and when it turned out that we had the problem, I happened to be the EECOM on the console. I don't think any other EECOM would have recognised that random pattern. Our simulators did not train us for it, but I saw it through the procedural screwup. Although the test happened a year before, that pattern was etched in my mind, and I am talking about a pattern of thirty or forty parameters. Instead of reading zeros, one would read six point something, another read eight point something, which were nonsense numbers for a 28 volt power system."

Aaron quickly called Capcom Jerry Carr on the voice loop to tell the spacecraft, "Flight, try SCE to Aux." In the spacecraft Bean heard Carr's instruction, found the Signal Condition Equipment switch, reached across to flip it down to "Auxiliary" which selected an alternate power supply, and order was restored to the television screens.

Aaron recounts, "We now got back live telemetry that was representative of the actual readouts on the spacecraft. We then realised that the fuel cells, the main power source, had been kicked off the line, all three of them, and the whole spacecraft was now being powered by the emergency re-entry batteries in the Command Module, which worked on a lower voltage. They were never designed to carry the full load of the Command and Service Module in a launch configuration. The next call I made was to reset the fuel cells and the voltage was returned to normal.

"I felt quite relieved just to get those guys into low Earth orbit, but I will never forget what Chris Kraft said to me that day, he said, 'Young man, don't feel like we have to go to the Moon today, but on the other hand if you and the other systems people here can quickly check this vehicle out and you feel comfortable with how to do that then we're okay to go, but don't feel you have to be pressured to go to the Moon today after what happened. We don't *have* to go to the Moon today.'

"We then dreamed up a way to do a full vehicle system checkout by improvising and cutting and pasting some of the crew procedures that they already had."

Nothing serious seemed to have happened, so while still hurtling ever faster up into space, the crew had restored all the systems except the inertial guidance system, and that was set by the 32 minute mark as they shot into the darkness over Africa.

There was some concern that the lightning may have damaged the parachute system in the nose of the Command Module or affected some of the Lunar Module systems at launch, particularly the highly sensitive diodes of the landing radar. With all systems apparently working normally *Intrepid* homed in to a pinpoint landing on the target, Snowman Crater and the Surveyor III spacecraft, 2,029 kilometres west of the Apollo 11 landing site.

As a panorama of the landing area spread in the window before him, all Conrad could see was a jumbled mass of similar shadows and craters. How could they possibly pick out a particular crater in the time available? Remembering the trouble the experts had locating the Apollo 11 landing point, Conrad felt apprehensive about finding a speck, the Surveyor spacecraft and its particular crater, buried among these thousands of lookalikes.

However their navigation was so accurate the automatic controls were taking them straight to the target area. When Conrad lined up the figures from the computer in the window he recognised the familiar shape of Snowman Crater coming into view. After taking over Program 66 manual control at 122 metres Conrad found he had to sidestep the Surveyor crater: "Hey, there it is. Son of a gun, right down the middle of the road. Hey, it started right for the centre of the crater. Look out there. I can't believe it … amazing … fantastic …," an incredulous Conrad remembered how he had asked trajectory specialist Dave Reed to target *Intrepid* for the middle of the crater, not really believing he could do it. Apollo 12 used a new computer program called a Lear Processor to minimise navigational errors using the three big tracking stations on Earth to correct *Intrepid*'s course, or it would have overshot the target by 1,277 metres.[46]

Conrad told Bean, "I gotta get over to my right," and searched for a clear area just beyond Snowman Crater until at about 30 metres the rocket exhaust kicked up a raging dust storm and Conrad lost sight of the lurain under the shooting bright streaks of dust blasting away from under their feet. Eyes glued to the instrument panel, occasionally flicking to look out the window, he had no idea whether there were threatening craters or boulders below, or not. The blue light lit up; Bean announced, "Contact light," and Conrad shut down the rocket motor. They dropped vertically to land with a solid thump about 6 metres from the edge of the Surveyor crater at 12:54 am on 19 November.

> Conrad: "I think I did something I said I'd never do. I believe I shut that beauty off in the air before touchdown."
> Capcom Jerry Carr in Houston: "Shame on you!"
> Conrad: "Well, I was on the gauges. That's the only way I could see where I was going. I saw that blue contact light and I shut that baby down and we just hit from about 6 feet (1.8 m)."
> Carr: "Roger. Break Pete. The Air Force guys say that's a typical Navy landing!"
> Conrad: "It's a good thing we levelled off high and came down because, I sure couldn't see what was underneath us once I got into that dust."

Gordon orbiting in *Yankee Clipper* 96 kilometres above, searched through a 28 power telescope and spotted a speck of light with a shadow, then another speck nearby, about three hours after they landed. He said excitedly, "I have ... I have Intrepid! I have Intrepid! The Intrepid is just on the left shoulder of Snowman ... I see the Surveyor! I see the Surveyor!"

"I can't wait to get outside – these rocks have been waiting four and a half billion years for us to come and grab them!" called an impatient Conrad as they worked their way through the essential housekeeping procedures. Five-and-a-half hours later Conrad emerged through the hatch and leapt onto the Lunar Module's footpad with both feet. "Whoopee! Man, that may have been a small step for Neil, but it's a long one for me!" he chuckled as he began to look around. Nobody remembers second, so his first words were said voluntarily to win a bet with an Italian journalist and to prove that Armstrong had not been pressured what to say by government officials. Then, "You'll never believe it. Guess what I see sitting on the side of the crater – the old Surveyor." The high spirited, exuberant Apollo 12 lunar excursions were a welcome contrast to the formal, tension filled, Apollo 11 lunar walk.

They had landed a mere 183 metres from Surveyor III, launched from Earth 31 months before. Their visit to it would have to wait for the next day, though, as the first task was to lay out all the equipment for the science experiments, the first ALSEP (Apollo Lunar Surface Experiments Package).

Conrad recalls, "And the dust! Dust got into everything. You walked in a pair of little dust clouds kicked up around your feet. We were concerned about getting dust into the working parts of our spacesuits and the Lunar Module, so we elected to remain in our suits between our two EVA's"

> Bean to Conrad: "Boy, you sure lean forward."
> Conrad to Bean: "... don't think you're gonna steam around here quite as fast as you thought you were."

Odyssey then plunged into the atmosphere to be engulfed in a streaming firestorm, a fireball streaking across the sky. All communications with the spacecraft were cut off.

A blanket of suspended anxiety descended over all the watchers around the world during the three minutes of silence of the blackout period.

The seconds flicked away with no response from the spacecraft. Unable to do any more for the mission now, the Houston Flight Controllers could only watch the recovery forces at work on their large television screens and listen for the spacecraft to respond to their Capcom.

The tension built up … a minute after the expected time and still no sight or sound of *Odyssey* – the cameras stared at a vacant sky, the speakers just hissed static.

Joe Kerwin called out from Houston, "*Odyssey*, Houston standing by."

Suddenly Swigert's voice filled the airwaves over the Pacific, "OK, Joe!" and soon three healthy parachutes could be seen. Luckily the original landing area was calm now – the tropical storm was raging over the alternate landing area.

Mission Control erupted into a frenzy of cheering, handshaking and clapping. John Aaron remembers, "Since I had designed the re-entry sequence, Kranz put me on the console as it came in and that worked out fine, but that sure was something when they made it through the blackout and out came the chutes. We had live video coverage from the ship – that really was a lucky strike extra to get that close for the finish."

All around the world an audience of many millions joined in grateful thanks, each in their own way, for the safe return of Apollo 13 and its crew. At 12:07 pm Houston time on Friday 17 April, the parachutes dunked *Odyssey* into the Pacific Ocean 6.4 kilometres from the USS *Iwo Jima*, and the crew were greeted by cheering sailors, a brass band, and Rear Admiral Donald Davis with, "We're glad you made it, boys." Nine doctors checked them out to be in reasonable shape considering their ordeal, except for a urinary tract infection for Haise, brought on by not drinking enough fluids, which allowed the toxins to build up. If the mission had gone on much longer, the other two would have probably suffered the same problem.

Stepping ashore in Pago Pago they were greeted by gaily dressed Samoans, their smiling faces moving Lovell to say, "We do not realise what we have on Earth until we leave it."

Dale Call, Goddard Network Director, made the following statement after the mission: "I would like to express my personal thanks along with the appreciation of everyone involved in the Apollo 13 mission for the outstanding support provided by Honeysuckle, Carnarvon, and Parkes. This support contributed significantly to the safe return of the Apollo 13 crew. I would especially like to single out those responsible for bringing up the Parkes antenna and associated data systems in record time. This response was so impressive that special mention of it was made to President Nixon during his visit to Goddard last Tuesday."[49]

President Nixon addressed a message to Australian Prime Minister John Gorton:

Dear Mr Prime Minister,

On behalf of the people of the United States I wish to express to you and to the people of Australia my deep appreciation for your nation's assistance in the successful recovery of the Apollo XIII astronauts.

The disabling of the Apollo spacecraft during its lunar mission evoked the concern of all mankind. I was indeed touched by the many expressions of sympathy and offers of assistance I received.

The safe recovery of the astronauts, for which we are all profoundly thankful, in no way lessens the gratitude of the Government and people of the United States for your nation's immediate response to our need for assistance.

Please convey my personal thanks to all of your people who worked so hard to maintain our communications with the weakened Apollo XIII spacecraft as it returned to Earth. Their involvement in the Apollo XIII recovery was but another instance of the close cooperation-operation and warm friendship that exists between our countries.

Sincerely,
Richard Nixon

For a brief moment Apollo 13 put the hassles of money and budgets and politics aside as people followed the progress of a mission where sheer guts and determination, teamwork and comradeship, ingenuity and skill brought the crew safely home. No doubt luck was a large factor in the equation. A triskaidekaphobic person would freak out at this list of thirteens – it was the 13th Apollo mission, launched at 13:13 hours spacecraft time, the explosion occurred on 13 April, 13:08 hours Honeysuckle Creek time, with 13 nations offering to provide rescue ships or aircraft. And stretching credibility a bit the astronauts first names of James, Fred, and Jack add up to 13 letters, the launch date of 11/4/70 add up to 13 from pad 39 which is 3 × 13. Even German Measles has 13 letters. Not surprisingly the Horoscope for Aquarius from the Houston Post of 13 April 1970 said "Do surprises turn you on? Then this is the day for the unexpected."

Was Apollo 13 good luck or bad luck? Probably good luck because it brings up a lot of "What ifs …?" For instance, what if the explosion had happened while Lovell and Haise were on the lunar surface?

Lovell said, "To get Apollo 13 home required a lot of innovation. Most of the material written about our mission described the ground-based activities, however, I would be remiss not to state that it really was the teamwork between the ground and the flight crew that resulted in a successful return. Some people would call the Apollo 13 mission a $375 million failure. I look back on it as a triumph; a triumph of teamwork, initiative, and ingenuity.

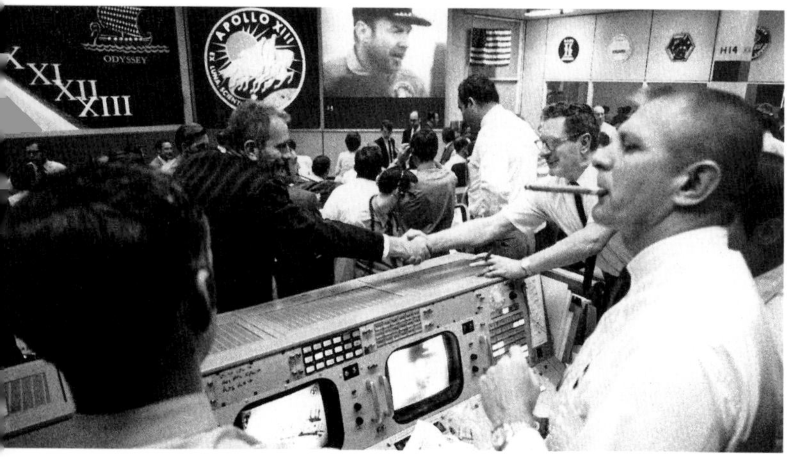

Figure 6.16. "Failure is not an option." Celebrating a successful landing, Gene Kranz on the right and Glynn Lunny on the left enjoy watching the handshaking and back slapping in Mission Control as the large television screen shows Apollo 13 Commander James Lovell arriving safely on board the USS *Iwo Jima*. Behind them Deke Slayton on the left shakes hands with Chester Lee, Mission Director from NASA Headquarters. This mission more than any other highlighted the need for real time support from the ground.

"Nobody believes me, but during this six-day odyssey we had no idea what an impression Apollo 13 made on the people of Earth. We never dreamed a billion people were following us on television and radio, and reading about us in banner headlines of every newspaper published. We still missed the point on board the carrier *Iwo Jima* which picked us up because the sailors had been as remote from the

Figure 6.17. On board the carrier *Iwo Jima* heading for Pago Pago Jim Lovell catches up with the news Apollo 13 was creating.

media as we were. Only when we reached Honolulu did we comprehend our impact."[17]

For fun, Grumman, the builders of the Lunar Module, sent a bill for $400,000 to North American Rockwell for towing the Command and Service Module 482,800 kilometres back home!

Chris Kraft says, "I think Apollo 13 was a classic example of what the ground flight operations was all about. It proved that the ground was worth while. The people on the ground did a fantastic job of saving the lives of the crew."

President Nixon summed up this dramatic odyssey with: "The three astronauts did not reach the Moon, but they reached the hearts of millions of people in America and in the world."

1970 **The Russians Land a Mobile Robot on the Moon**

The Russians, thwarted with landing a man on the Moon, were trying various robot missions. While there was a ten-month break in the Apollo lunar missions, the Luna 17 mission was launched on 10 November 1970, which successfully landed a 754 kilogram remote-controlled rover they called *Lunokhod* in the Sea of Rains on 17 November. By its fourth lunar night on 20 February 1971 it had travelled 4.8 kilometres and conducted experiments which included a "thumper" to compact the soil to test it for mechanical strength.

The crude television system, using only one frame every three seconds, caused the ground scientists great difficulty in the early days of the exercise with scaling and fast driving. Once, the scientists ordered the vehicle to move toward what appeared to be a large boulder about 100 metres away. The "boulder" disappeared in the next frame, and they discovered it had only been a small rock a metre or so away. Lunokhod 1 lasted for 10 months and crawled over 10.5 kilometres of Sinus Iridum. It did not appear to add much new information on the Moon.

1971 **Apollo 14 Makes it Safely to the Moon and Back**

Roosa named the Command Module *Kitty Hawk* in memory of the first powered flight by the Wright Brothers, and Mitchell called the LM *Antares* after the star in the Constellation *Scorpius*, as it was a significant mark visible through their window as they

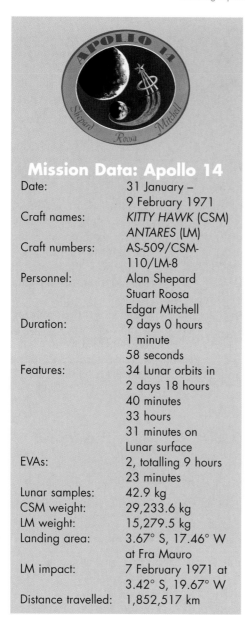

Mission Data: Apollo 14

Date:	31 January – 9 February 1971
Craft names:	*KITTY HAWK* (CSM) *ANTARES* (LM)
Craft numbers:	AS-509/CSM-110/LM-8
Personnel:	Alan Shepard Stuart Roosa Edgar Mitchell
Duration:	9 days 0 hours 1 minute 58 seconds
Features:	34 Lunar orbits in 2 days 18 hours 40 minutes 33 hours 31 minutes on Lunar surface
EVAs:	2, totalling 9 hours 23 minutes
Lunar samples:	42.9 kg
CSM weight:	29,233.6 kg
LM weight:	15,279.5 kg
Landing area:	3.67° S, 17.46° W at Fra Mauro
LM impact:	7 February 1971 at 3.42° S, 19.67° W
Distance travelled:	1,852,517 km

flew down to the lunar surface. At 47 years, Mission Commander Alan Shepard was the oldest man, and the only one of the original Seven Mercury astronauts, to walk on the Moon.

Eugene Cernan recounts, "I was Commander of the back-up crew for Apollo 14 – we had the only back up crew patch in the entire space programme. After ours, back-up crew patches were outlawed. Ron Evans, Joe Engle and I called ourselves the 'First Team'. We called the Prime Crew: Shepard, The Old Man; Mitchell, The Fat Man; and Roosa, the Cute Little Redhead. We embodied them all in one character, the coyote from the Roadrunner cartoon series. We had him coming out from the Earth with a big trail of fire behind, and he's looking at the Moon. He's all red, he's got a beard, and he's got a big fat belly – The Old Man, The Fat Man and The Cute Little Redhead! And there on the Moon is the Roadrunner – the 'Beep Beep' guy. The Roadrunner is sitting there with a big silk scarf round his neck, his legs are crossed, and he's leaning on an American flag. And this coyote is looking at the Moon, at this First Team, being back up crew, who got to the Moon before they did! We put about 20 or 30 of these patches inside both their spacecraft so every time they opened a locker in zero gravity one of these patches would float out."

Because the scientists had given Fra Mauro a high priority, it was re-assigned from the Apollo 13 mission. The first two landings had been on easy, flat territory, but Fra Mauro was the first of more challenging landing sites, a range of rugged mounds 177 kilometres to the east of the Apollo 12 landing site. A legacy from Apollo 13 were changes to the spacecraft to try and prevent another explosive, cliff hanging mission. This time there were three oxygen tanks, instead of two, the third isolated, and a new spare 400-ampere battery to carry the mission from any point. However this mission came up with new twists to keep the crews and Flight Controllers on their toes, and to remind everyone once again these spaceflights are never a routine operation.

The Spanish Apollo tracking station is about 60 kilometres north west of Madrid, and the staff travelled through secondary roads and little villages to get to work. On Apollo 14's launch day a convoy of about 30 station cars containing a mixed group of Spanish

Figure 6.18. The Lunar Module assembly line. The Apollo 15 Lunar Module on the left heads for the altitude test chamber, while the Apollo 14 LM on the right has its landing gear installed.

and American technical staff reached the village of Valdemorillo and ran into a religious festival. The holiday crowds completely blocked the road. Station staff member José Grandela remembers, "The only perceptible sound came from a music band while more than a thousand persons accompanied them in respectful silence. The traffic police were stopping all the cars until the procession was over. Most of us went out of the cars to watch the religious act but we were recognised by some of the assistants to the ceremony. A rumour started to go through the crowd while everybody glanced at us and pointed their fingers at us.

"We then noticed a group of policemen having difficulty getting to us through the mass of people. The officer in charge asked us, 'I have been told that you belong to the American base and that the astronauts on the Moon could be at serious risk if you don't reach the base on time, so we are going to stop the ceremony and allow you to go through the crowd.'

"I don't remember that any of us replied a single word – we were just astonished. We got back into our cars and drove at walking speed, physically touching the peasants while some comments, almost in a whisper, reached us, 'They are the Men of the Moon … they are the Men of the Moon.' Once we passed the last house of the village we burst into laughter and didn't stop until we reached the station."

After departing from Kennedy Space Center's Launch Pad 39A at 4:03:02 pm EST the astronauts followed the normal routine of extracting the Lunar Module from its launch housing. As Stu Roosa skillfully brought the Command Module in to the Lunar Module docking cone, the astronauts confidently waited for the thud of the latches biting, and green light to confirm a hard dock. To their surprise, even though they appeared to have made solid contact – there were no thuds from the latches and no green light! They had bounced off! It was unbelievable. This was the first time the American's had a docking failure at their first attempt.

Roosa called in, "Houston, we've failed to secure a dock."

A surprised Houston responded with, "Roger, Kitty Hawk. You've got a *go* for another attempt."

The Flight Controllers sat up and began to think about possible causes and how to overcome this new development. They looked around for the specialist engineers, and the engineers began to look for their ground replicas and procedures. If there was something wrong and they were unable to dock, this would be the end of the lunar landing part of the mission, and possibly all further Apollo missions as there were already authoritative voices calling for an end to any more lunar flights in case tragedy struck – quit while ahead! Then to their dismay they heard Roosa's frustrated voice after the second attempt. "Houston – we do not have a dock. We're going to pull back and give this some thought."

At the critical moment Mission Control discovered the replica docking system could not be found. Director of Flight Operations Chris Kraft explains, "Previously we'd always had a docking probe and drogue available

Figure 6.19. In the Kennedy Space Center's Firing Room No. 2 it was 9 am on 25 January 1971. Test Supervisor Charles Henschel pushed the switch to start the clock to launch Apollo 14, planned for six days later. The clock began to count down to launch from 102 hours, and a vast army of technologists in America and around the world hooked on to its beat to get the third manned landing on the Moon under way.

Figure 6.20. Dr George Low (left), Deputy Administrator of NASA, watched the Apollo 14 launch with rocket genius Dr Wernher von Braun in the Launch Control Center's No. 2 Firing Room.

at the Control Center, as well as experts on the system, but now there were frantic calls for assistance and the absent docking system had to be hurriedly located to understand what might be going on thousands of miles out in space."[17]

Three times over the next hour they tried docking without success, while the replica in Mission Control never failed. "It's possible there is some dirt, or debris, in the latches," suggested an engineer, and as fuel was beginning to run down, they decided to try a "do or die" attempt by coming in fast, ramming the probe and drogue together and hitting the switch for a hard dock, bypassing the normal procedure of a soft dock first. Hopefully any possible foreign matter would get dislodged.

Roosa: "Houston – we're going in …"
Houston: "Good luck, Kitty Hawk …"

Houston could only stand by and listen. Out in space Roosa glanced at Shepard and saw the Icy Commander – angry. "Stu, just forget about trying to conserve fuel. This time … juice it!" Shepard growled at him.

The three men held their breath as Roosa gunned his ship and the Command and Service Module obediently leapt forward and slammed accurately into the Lunar Module. The crew steeled themselves for the rebound but the latches dropped into place and a green capture light glared at them from the control panel.

"Got it!" yelled the crew in unison.

Now Smilin' Al turned from his instrument panel and quietly announced, "We have a hard dock."

Roosa keyed his transmit button, and tried not to shout in glee, "Houston, we have a hard dock."

Another crisis in the Apollo Program passed into history and the mission continued to follow the flight plan until they went into orbit around the Moon and it was time to land. Following normal procedures they initiated a computer practice run to land. The computer program started all right, but then without warning, flung itself into an abort mode to return back to *Kitty Hawk* without landing.

Shepard called out, "Hey, Houston, our abort program has kicked in!"

Every try produced the same result, and every check could find no errors. The lunar landing was put on hold while ground trials and evaluations finally found the problem to be a faulty abort switch, so they yanked computer specialist Donald Eyles out of bed in Massachusetts to write a new program to accommodate this faulty switch, and transmitted it up through the tracking stations to the spacecraft circling the Moon. Shepard, itching to be doing something but only able to wait, anxiously watched Ed Mitchell load and check out the computer, then called with relief, "Houston – we've got it. We're commencing with the descent program."

"*Antares*, you have a *go*," replied the Houston Capcom.

It was close. There were fifteen minutes left. Fifteen minutes before having to abort and return to *Kitty Hawk* without landing. The next fright came as they approached the surface. The landing radar refused to lock on initially due to the system switching to a low range scale and if it did not find the target by 3,048 metres altitude, mission rules specified an abort. Houston were working on the problem and Capcom Fred Haise radioed up, "We'd like you to cycle the Landing Radar breaker."

Shepard pulled the circuit breaker out and pushed it back. "OK, it's cycled."

Within seconds the caution lights went out and there was good data being displayed. Shepard and Mitchell went on to execute the most accurate landing of the Apollo Moon landings, putting *Antares* down only 53 metres northeast of the planned landing spot at 3:18 am on 5 February. Shepard is reputed to have dropped it short on purpose as it was in the direction they were to walk first, and it would save them some walking, but he wrote, "The landing site was rougher on direct observation than the photos had been able to show. So I looked for a smoother area, found one, and landed there.

"Ed and I worked on the surface for 4 hours and 50 minutes during our first EVA; after the return to *Antares*, a long rest period, and then re-suiting, we began the second EVA. This time we had the MET – Modularised Equipment Transporter, although we called it the lunar rickshaw – to carry tools, cameras, and samples so we could work more effectively and bring back a larger quantity of samples. We covered a distance of about two miles and collected many samples during $4\frac{1}{2}$ hours on the surface in the second EVA. I also threw a makeshift javelin and hit a couple of golf shots."[17]

Shepard's prosaic description of the second walk did not tell the real story. The excursion turned out to be a nightmare of navigation problems and obstructive terrain. Leaving *Antares* just before 3:00 am they headed east, into the rising Sun. Looking for Cone Crater, the major objective of their mission, they waddled about ground littered with rocks that were probably scattered when the crater was formed. The crystal clear sharpness of a world devoid of atmosphere distorted their perception of depth and distance. Unable to find any feature in the stark colourless scene to guide them, they struggled to locate the direction of the crater. As they neared what they believed to be the crater's sides the boulders grew bigger, unexpected gullies slowed them up, optical illusions plagued their judgment, and they had to slog through ground covered with rubble and broken rocks, some as big as a small house. They found it was easier to carry the MET up these slopes of rubble. Some slopes were like slip sliding up a dune of dry sand.

Figure 6.21. A dramatic picture of *Antares* sitting with an 8° tilt on the lunar surface at Fra Mauro only 26 metres from the chosen target point. On the left of the picture the lower slope of Cone Crater can be seen. The gold foil wrapped around the descent stage was needed for temperature control.

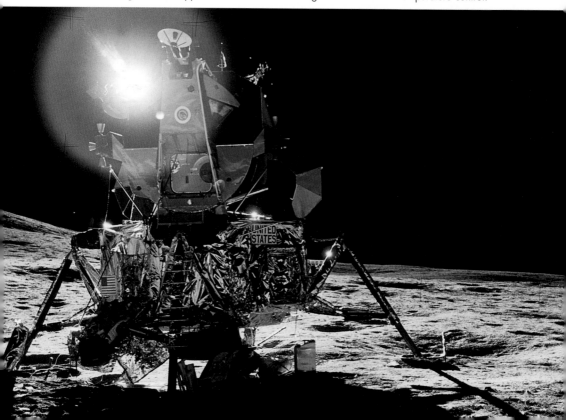

nutrient content to the original food. The food was quickly frozen down to −10°C and packed in a vacuum chamber for slowly reheating for use later. This improved the fare and the Apollo 15 team were the first crew to eat all their rations. They were also the first astronauts to eat solid food bars while working on the lunar surface

Irwin says, "When it was announced that the crew was eating, everybody thought that we must be relaxing. Believe me, it was not like that at all. First you've got to find the meal for that particular time of that particular day. Of course we had a stowage map which told us where everything was when we left the Earth. All food containers were labelled A, B, C, D, E, and all meals were colour coded. Mine were blue, Dave's red, and Al's white.

"So if a meal floated by and you identified it as being red, you could say, 'Hey, Dave, you've just lost your entrée.'

"Actually, we had to improvise a way of managing this flying circus – it worked out best if I prepared the food. Three people can't use one water gun at a time to mix different amounts of hot or cold water with ingredients of eighteen bags. I read the directions – six ounces, eight ounces, hot or cold. And after I had shot the water into the bag, I'd wait five or ten minutes for the bulk to absorb the water. Or, I'd float it out and let the customer age his own bag. Anyway you did it, this had to be the most unusual small restaurant in the world or out of it.

"We had Velcro all over the spacecraft and on the meal packages, so to keep track of things we'd stick our dinner on the wall, course by course. If you nudged the meat course accidentally, it would take off, and you would have to float after it or get the help of a buddy downfield."[51]

After each meal they had to log down everything they ate, or didn't eat, and advise Houston, who kept track of their energy levels.

Irwin continues, "We brushed our teeth after each meal. I had taken a razor because my face starts itching when I grow a beard. But Dave and Al hadn't even brought razors, so I decided that if they weren't going to shave, neither would I.

"Bathing was another chore. We would dampen a washcloth and clean ourselves all over as well as we could. For some reason the other guys didn't bring any soap. Fortunately I took a bar of sweet smelling soap along. It was the high point of the day just to take out the soap from the container and let the scent waft around the spacecraft. It almost made us feel clean."[51]

The astronauts tried the first scientific experiment of the trip by putting shades over their eyes and looking for strange orange flashes reported from earlier flights. They reported seeing 61 flashes. "They look like flashbulbs popping in a darkened arena," commented Scott. They are believed to be high-energy cosmic rays impacting the eye's retina, or perhaps the brain's optical centre.

Once they were safely orbiting the Moon, Scott was fascinated by the scenes of the moonscapes 96.5 kilometres below them, the dark side away from the Sun and Earth, with earthlight brighter than the moonlight on Earth. They could easily see the mountains and crater rims glowing from the reflected light from the Earth. But it was the sunrises every two hours that impressed Scott the most. "First of all these wispy streamers of light from the Sun's corona appeared above the lunar horizon, then the Sun simply exploded over the horizon like a visual thunderclap, and within a second we were blinded by its bright light flooding the cabin."

Just before landing they dropped into a 15×72 kilometres orbit. As they zoomed low over the lunar surface, Irwin realised that there were mountains higher than they were. "You look out on the horizon and you see these high peaks and you are just skimming along. Now you really know you are moving fast. You are travelling about 5,000 feet per second, that's Mach 5 or 3,000 miles per hour (4,828 kilometres per hour). Your orbit is defined; you can't dodge anything. You don't have control over the vehicle, and if you did you probably couldn't react fast enough. You just assume that Houston knows where the mountains are and how high they are. But you see the high mountains on the horizon and you move towards them very fast. You wonder if you are going to clear them.

"The face of the Moon is beautiful in a stark, awesome, barren way. It is all ochre's, tans, golds, whites, grays, browns – no greens, no blues. We were hanging loose, coming to the burn itself. We were going to land on the surface. Dave was doing most of the hand-control action, but in the main we were telling the computer what to do."[51]

> Mitchell: "Fifteen – does it look like you are going to clear the mountain range ahead?"
>
> Scott: "We've all got our eyes closed; we're pulling our feet up."

Scott concentrated hard on bringing Falcon down to the final stages of this tricky landing, a target at the bottom of a basin hemmed in on three sides by mountains, and on the fourth by a deep gorge. Approaching a height of 2,438 metres above the valley surface, they were flying horizontally, looking straight up before pitching forward at 1,830 metres to be able to see where they were. Both astronauts were suddenly startled to see the white flank of a mountain sliding past *above* them out of the left window! It was Mount Hadley Delta soaring up 3,962 metres from the plains. The simulator had never shown them this; was Houston aware how close it was? What if they had been off course?

Scott looked as far forward as he could but still couldn't see any sign of Hadley Rille. "I looked out the window and could see Mount Hadley Delta. We seemed to be floating across Hadley Delta and my impression at the time was that we were way long because I could see the mountain out the window and we were still probably 10,000 to 11,000 feet (3,353 metres) high. I couldn't see the Rille out the forward corner of the window, which you could on the simulator – out the left forward corner."

Mitchell spoke to the astronauts. "*Falcon*, Houston. We expect you may be a little south of the site … maybe … 3,000 feet."

When *Falcon* pitched over on time all Scott saw was a featureless plain below them. He was looking for Index Crater, where they were supposed to land. "I couldn't convince myself that I saw Index Crater anywhere. I saw, as I remember, a couple of shadowed craters, but not nearly as many as we were accustomed to seeing in the simulator. Once I realised that we were not heading for the exact landing site, and I didn't have a good location relative to Index Crater, I picked what I thought was a reasonably smooth area and headed directly for that."

Then in the distance ahead he could make out Hadley Rille, so Scott manually brought the Lunar Module down to where he thought the planned landing should be. "At about 60 feet (18 metres) the dust came up at us and I lost sight of everything and concentrated on Jim's calls. I hoped there were no boulders or craters under us … we were dropping blind – then Jim called, 'Contact' and I shut the motor off."

Irwin agrees. "The light came on. I called, 'Contact!' and Dave immediately pressed the button to shut the engine – then we fell. We hit. We hit hard. I said, 'BAM!' but it was reported in some of the press accounts as 'damn'. It was the hardest landing I had ever been in. Then we pitched up and rolled off to the side. It was a tremendous impact with a pitching and rolling motion. Everything rocked around and I thought all the gear was going to fall off. I was sure something was broken and we might have to go into one of those abort situations. If you pass 45 degrees and are still moving, you have to abort. If the Lunar Module turns over on its side, you can't get back from the Moon."[51]

At 5:16 pm on 30 July *Falcon* landed on the lunar surface with the hardest of the Apollo landings at two metres per second. Scott had by far the heaviest spacecraft to that date with the first Lunar Rover aboard. He was also very quick to switch the engine off as he wanted to make sure the engine was off before the bell housing, which was longer than the earlier models, could contact the surface.

"Okay, Houston. The *Falcon* is on the Plain at Hadley," Scott advised Mission Control.

They waited for Mission Control to look at all the systems and give them a *go* for staying on the Moon, and when it came they began to power the spacecraft down.

Irwin recalls, "Dave and I pounded each other on the shoulder, feeling real relief and gratitude. We had made it."

Falcon had landed on the edge of Mare Imbrium, which stretched across the surface of the Moon for at least 1,046 kilometres to the west. The three legs of the Lunar Module had straddled the rim of a crater with a tilt back of 6.9° and a lean of 8.6° to the south. The descent engine bell had been damaged a bit, probably from pressure build-up on landing.

At 7:16 pm Scott climbed up on the engine cover, opened the top hatch and gazed out on the lunarscape. Stark, white craters scarred the soft beige of the flowing lurain. Dark lines ran around the foot of the mountains. Like a magnet his gaze was drawn up to the blue and white Earth glowing in the impossibly black sky – the only colour in the whole scene before him. He spent thirty three minutes just studying, photographing, and reporting his observations back to Mission Control in Houston. NASA Geophysicist Robin Brett said his descriptions were as good as a professional geologist, many agreeing it was the best geological description by an astronaut on the Moon.

Scott says, "The incredible variety of landforms in this restricted area (on the Moon the horizon lies a scant mile and a half from the viewer) fills me with pleasant surprise. To the south an 11,000-foot (3,353 metres) ridge rises above the bleak plain. To the east stretch the hulking heights of an even higher summit. On the west a winding gorge plunges to depths of more than 1,000 feet (305 metres). Dominating the north eastern horizon, a great mountain stands in noble splendour almost three miles above us.

"Ours is the first expedition to land amid lunar mountains. Never quickened by life, never assailed by wind and rain, they loom still and serene, a tableau of forever. Their majesty overwhelms me."[52]

Feeling as if they were intruders in an eternal wilderness, they closed the hatch, repressurised the Lunar Module, and turned in to sleep. Irwin recalls, "Dave was sleeping fore and aft, and I was athwart ship, with my hammock slung under his. I noticed that my hammock was bowed out a little bit and my feet were sort of dangling off.

Figure 6.26. At Mission Control on the seventh day, Capcom Scientist–Astronaut Dr Joe Allen followed the route of the second excursion with a map of the Hadley Appenine area as Scott and Irwin drive their Rover south towards Mount Hadley Delta.

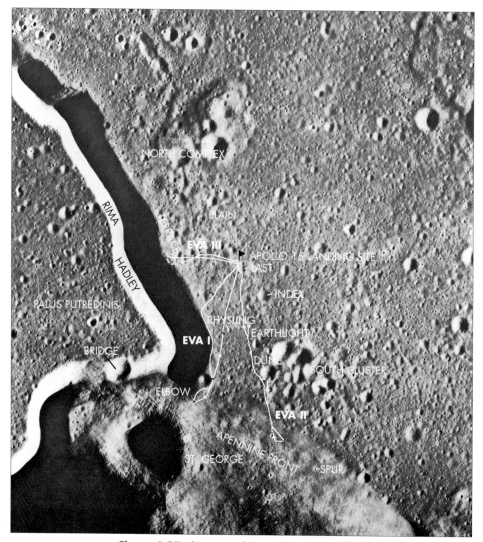

Figure 6.27. The routes of the Apollo 15 excursions.

It was noisy in the Lunar Module with the pumps and fans running, something like sleeping in a boiler room. But, man, it was comfortable sleeping! Those hammocks felt like water beds, and we were light as a feather. The first night's sleep was the best I had the three nights we were there."[51]

The next day, following Scott down the ladder, Irwin bounced down to the surface, kicked up a spray of black moondust, and looking to the south, exclaimed, "Oh, boy, it's beautiful out here. It reminds me of Sun Valley." The Apennine Mountains looked familiar to him, rounded and treeless. He thought they looked like excellent ski slopes. After five days of being cooped up in the spacecraft, they felt the relief and pleasure of being able to move around again, the freedom of room to run in. They felt it was like walking on a trampoline, the same bouncy feeling, and falling down was quite different to Earth – you seemed to go down in slow motion with only a light impact that they felt would never cause any harm.

The astronauts tugged the two "D" rings to release the rover but it refused to drop down. It took them a half an hour's struggle to drop the brand new $12.9 million bundle of tightly packed wheels, seats, armrests, footrests, fenders, batteries, motors, consoles, cables, and hinges, that made up the lunar rover onto the surface. After running through the checklist, Scott called Joe Allen at Houston. "I don't have any front steering, Joe." After more checks, "Still no forward steering, Joe."

After physically trying to turn the wheels, they gave up and drove off. Apart from the front wheel drive failing, the little car darted happily about before the explorers headed south on their first excursion.

Figure 6.28. Instrument and Communications Systems Officers (INCOs) Edward Fendell (left) and Granvil Fennington were responsible for remote control of the Lunar Rover's television camera. As it took 4.8 seconds for the signal to arrive at the console the operator had to try to anticipate the action for camera movements by up to five seconds.

Scott relayed his experiences. "The steering is quite responsive, even with only the rear steering. I can manoeuvre pretty well with the thing. There is no accumulation of dirt in the wheels."

"Just like the owner's manual," responded Allen from Houston.

As they drove across the lurain, they found the wheel fenders kept the dust down, and seat belts were needed as the rover often became airborne as it flew over the surface with a pitching motion rather like a cross between a boat in a lumpy sea and riding a horse. One of the fears had been that as the rover sped across the surface it would vanish into a cloud of dust thick enough to block the astronauts' view to see anything. Luckily this didn't happen, the wheel fenders keeping the dust down. With no air, there was no breeze blowing past.

The Commander was the main driver, sitting in the left seat, holding a "T" bar in his right hand, tilted slightly towards him to lessen fatigue. The bar gave full control of the rover – pulled to the left turns all four wheels for a left turn, right for right turn, pushing forward drives the vehicle forward, and pulling back can stop it in its own length. A switch on the base puts the rover in reverse, with speed controlled by pulling the T bar back. As it didn't have a rear view mirror, the astronauts usually preferred to just pick it up to turn it around as it only weighed 36 kilograms on the Moon.

Power was provided by a $\frac{1}{4}$ horsepower DC electric motor with a 80:1 reduction gear on each wheel, fed by two 36-volt batteries with silver–zinc plates in potassium hydroxide with the capacity to last twice the planned distances.[53]

On the first excursion, though the rover had quite sophisticated navigation facilities, in the beginning they had some trouble locating their exact position but suddenly on the top of a ridge Scott called out, "Hey, there's the Rille," and around 6:00 am they pulled up beside the 366 metre deep chasm, surprised at its vast size. It sloped down to the 198 metres wide floor on their side, but on the other side, about 1.6 kilometres away, it dropped steeply and was strewn with boulders. Right at the bottom were two house-sized boulders.

Scott found a smooth spot that sloped steeply into a gully. "It looks like we could drive down to the bottom over here on this side, doesn't it? Let's drive down there and sample some rocks."

Irwin was more cautious. "Dave, you are free to go ahead. I'll wait here for you." He felt he could not face the furore if they lost the rover in the Rille.

They followed the edge of the Rille around to Elbow Crater, 2.9 kilometres from the Lunar Module, and stopped short of the big St George Crater before heading back to the *Falcon*. They hadn't appreciated how high they had climbed until heading back. "Hang on!" yelled Irwin as they raced downhill.

Irwin was impressed with communications from the rover. "Communications between us and Dr Joseph Allen, our Capsule Communicator during EVAs on the lunar surface, were so clear that it was hard to believe we were really on the Moon. It was as if 'Little Joe' were sitting on one of those mountains talking to us. This seemed to bring us closer to home. Actually the radio signal (S-Band) suffered minimal loss going through space. It was sent from Honeysuckle Creek, Australia, our prime station."[51]

At the Lunar Module Irwin spread out the ALSEP science experiments, while Scott tried to drill a 3-metre core sample, but unfortunately it jammed in the extremely dense lurain. Later, it took the combined efforts of the two strong astronauts to work it out, eventually heaving it out with their shoulders. Analysis of the core on Earth identified forty-two layers of soil, the lowest layer half-a-billion years old.

After the first day's activities both astronauts were totally exhausted, both suffering almost unbearable pain in their fingers from working with the gloves. When they took their gloves off the perspiration just poured out. Before the mission Scott and Irwin wanted to try to get more feeling from their gloved fingers. At rest their gloves were designed to assume a half-open finger position, which meant they had to be forced to close over an object to hold it. So the astronauts had organised the arms of their spacesuits to be shortened so that the tips of their fingers just came in contact with the rubber finger ends of the gloves. After working like this for a while the pain from the fingertip pressure became excruciating, though later Irwin found cutting his fingernails eased the pressure. He was also unable to get to his suit drink water bag to release its contents, so never managed a drink during the whole seven hours of the excursion.

Figure 6.29. As the Prime station for Apollo 15, Honeysuckle Creek closely followed events on the lunar surface in the darkening twilight of the Australian bush.

The next "morning" the astronauts were roused up at 2:35 am to fix a leak of water from a broken filter before the second day's excursion in the rover took them across the plains, skirting Crescent and Dune Craters, to the Apennines. Mission Control had figured out a fix for the rover's front steering and the astronauts found it now worked brilliantly. As they approached Spur Crater on the flanks of the mountain they spotted an interesting looking boulder down a slope and stopped to investigate.

Irwin looked back and grunted, "Gonna be a bear to get back up there, y'know!" Then without warning the rover began to slip down in the powdery dust, one wheel rearing up in the air, and Scott leapt to grab it, conscious that it was nearly a 5 kilometres walk back in their suits if they lost their vehicle. He held it safe while Irwin looked at the boulder.

"Come and look at this," said Irwin, "This is the first green rock I've seen." Finding another he added, "I hope it's still green when we get home." It was – the samples were bright green glass beads formed as the soil was compressed by meteorite impact, later to add an important clue about the interior of the Moon.

They pushed on to Spur Crater where they parked the rover. Climbing out of the vehicle they found they were surrounded by a mine of geological treasures. White, light green, and brown rocks lay all around – then in the distance they spotted one piece of rock that looked different from the others.

Irwin recounts, "Finally we worked our way over to the white rock that had rivetted our attention

Figure 6.30. Inside, management at Honeysuckle Creek expected the staff to drink lots of coffee to keep the station running through the small hours. Paul Mullen reaches for a styrofoam cup as Mike Linney prepares another caffeine fix, while...

Figure 6.31. ... on the Moon, Scott and Irwin drove their new car around the mountains and rilles of the Moon. Irwin took this picture from the flank of St George Crater looking north along Hadley Rille showing Scott with the rover.

earlier. It was lifted up on a pedestal. The base was a dirty old rock covered with lots of dust that sat there by itself, almost like an outstretched hand. Sitting on top of it was a white rock almost free of dust. From four feet away I could see unique long crystals with parallel lines, forming striations. This was exactly the kind of rock we were looking for; it confirmed the suspicion that the mountains of the Moon were made from rock, lighter in colour and lighter in density."[51]

They were looking at the oldest piece of rock to be found by the Apollo missions.

Scott exclaimed, "Guess what we found! I think we found what we came for!" The strips of white crystals were what geologists call plagioclase twinning. "It's almost all plag – something close to anorthosite … it's almost all plag. …" They studied the 269-gram piece of original lunar crust before carefully labelling the sample 15415 and putting it in the collection.

Back on Earth this rock was dated at 4.15 billion years. As a comparison the oldest rock ever found on Earth so far is 3.3 billion years. It also turned out to be the oldest rock brought back from the Moon, so was christened the Genesis Rock by the media. They also found a 103 square kilometre field of cinder cones, the first positive sign of explosive volcanism on the Moon.

Standing at the bottom of Mount Hadley Delta, Scott looked up and remarked, "My, oh my! This is as big a mountain as I ever looked up." Half a mile up the mountainside, they looked back on a magnificent, though desolate, panorama of the basin, the tiny *Falcon*, the Rille, and the surrounding peaks.

Because of a shortage of time, the third excursion was drastically shortened to a visit to the rim of the Rille west of the Lunar Module, before preparing to depart. Years later the geologists said that the Apollo 15 samples were the most interesting and varied of all the missions.

Before leaving Scott then cancelled some stamps in front of the television camera for a first day of issue of 2 August 1971. A side issue of this episode created a lot of bad feeling later when it was discovered the astronauts had made special arrangements for selling 440 unauthorised stamped envelopes for their own financial gain, the scandal ending up before the Senate. That episode marked the end of the astronaut line for the crew of Apollo 15.

After cancelling the stamps, Scott and Irwin stood in front of the camera to demonstrate their "Galileo Experiment"; that all objects fall to the ground at the same speed in a vacuum. The original theory is actually attributed to Flemish mathematician Simon Stevin (1548–1620). In front of the television camera, Scott demonstrated this phenomenon by dropping a falcon feather from an Air Force mascot, and a geology hammer. They hit the lunar dust together in 1.3 seconds. The feather was lost when Irwin inadvertently trampled it into the dust, and he speculated on what sort of creature future explorers would imagine it would belong to.

This mission was the first time a camera could video the Lunar Module lifting off. Scott parked the rover about 90 metres from the Lunar Module, left the keys and the driving instructions on the seat, and checked the camera was looking at the Lunar Module for Ed Fendell (Captain Video), in Houston, who had remote control of the camera. For Apollo 15 the camera was left fixed looking at the Lunar Module because some lunar grit had seized the tilt gearing. Scott then knelt down to leave a stylised

Figure 6.32. The US Air Force song "Into the Wild Blue Yonder" burst down the voice channel as the ▶ ascent stage of *Falcon* rose into the black lunar sky. This picture sequence was taken by a camera on the parked Rover and recorded by Ed von Renouard at Honeysuckle Creek. This was the first time a lunar launch had been seen from Earth.

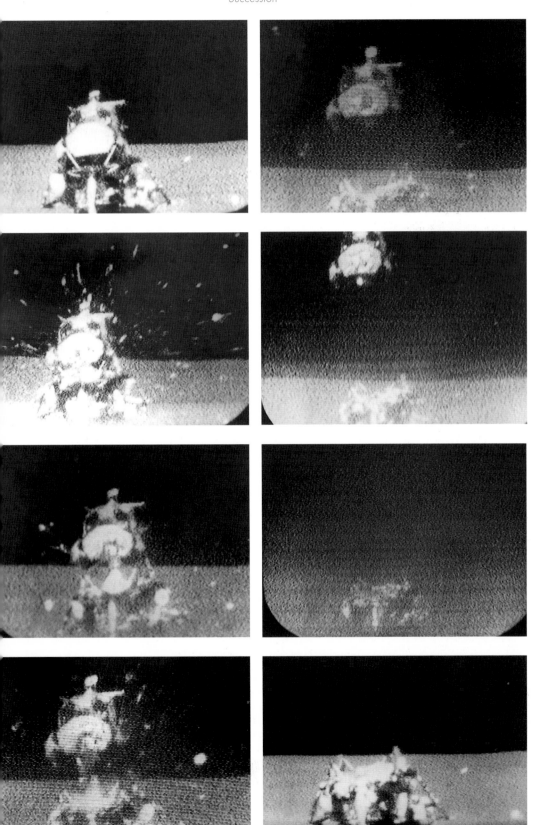

aluminium figurine of a fallen astronaut beside a plaque with the names of fourteen Russian and American men who had given their lives for spaceflight up to that time.

Lifting off at 12:11 on 2 August *Falcon* blasted the launch area and vanished rapidly out of the top of the picture, unexpectedly accompanied by a rousing version of the Air Force song "Off we go, into the wild blue yonder. ..." Scott had organised Worden to switch a tape recorder on one minute after lift-off, but it came bursting down the voice lines at ten seconds after to the surprise of all the flight controllers.

In later missions the camera followed the Lunar Module up. This, though, was very tricky, because it took 4.8 seconds for the picture to arrive at Fendell's monitor in Houston due to the distance and decoding of the signal. So to get the lift off he had to anticipate the launch by over 5 seconds and tilt the camera up to follow the Lunar Module by the scheduled launch time – not by the picture he could see on his monitor.

Signs of physical strain with the two astronauts became evident during the mission as they worked, particularly the hard time with the core drill. Both Scott and Irwin had spent long hot periods in spacesuits during training on Earth, drinking a solution that drained potassium from their systems. The day before launch Irwin had played a series of hard games of tennis and had became physically and emotionally tired and so dehydrated he needed three bottles of Gatorade to recover. He suspected this helped strain his

Figure 6.33. Piloted by Al Worden, the CSM *Endeavour* flying high above a sun-drenched Moon. The Scientific Instrument Module (SIM) in the Service Module that scanned and photographed the lunar surface can be seen in this photograph taken from the Lunar Module. An RCA laser altimeter and Fairchild mapping camera are in the section nearest the Command Module, with an Itec panoramic camera in the centre left. Spectrometers are in the section nearest the SPS engine bell.

heart and lower his potassium level before the mission. Watching their EKG monitors the doctors became aware of irregularities in both their hearts. Irwin, in particular, was showing signs of bigeminal rhythms (the two chambers of the heart try to contract at the same time sometimes caused by chronic fatigue) during the third EVA.

As the astronauts prepared to jettison the Lunar Module they had been up for 17 strenuous hours. Irwin's body sensors were again transmitting the presence of bigeminal rhythms. Small errors in judgment and memory lapses seemed to indicate ominous signs of excessive fatigue to the flight controllers on the ground. The suit and spacecraft pressure integrity checks both had to be repeated due to anomalies, then Scott thought they were going to collide with the Lunar Module if they followed the Mission Control procedures. Flight Director Kranz noted that he had never seen a spacecraft crew and ground crew so out of phase, "I was spooked just listening."

Unaware of the solicitous state of Mission Control the astronauts pressed on. Irwin remembers, "We really felt this separation; she was the best. Our Lunar Module just sat there in space as we backed away, then all of a sudden, for the first time in the flight, I felt really tired."[51]

Irwin told Scott and Worden he was going to lie down for a while, but back in Mission Control Deke Slayton had been advised of the doctor's concern for their health so called the spacecraft. "You guys have had a hard day, I suggest that you take a Seconal this evening and get a good night's sleep." As Mission Control still had not told the two astronauts of their health problems they did not take Slayton seriously; or the Seconal sleeping tablets. Scott did not feel they had reached their limits of endurance at any stage of the mission.

Endeavour and her crew continued orbiting the Moon for another two days, busy with experiments as well as mapping and photographing the lunar surface. During the Rev 74 they spring-ejected a Particle and Fields Satellite to help determine the lunar gravity field anomalies.

On 7 August, Apollo 15 returned back to Earth and the carrier USS *Okinawa* waiting in the Pacific, 288 nautical miles north of Pearl Harbor. Spotting the spacecraft swinging under only two parachutes, the third bunched up under its drogue, the recovery team radioed the spacecraft, "You have a streamed chute. Standby for a hard impact." With only two parachutes they landed with the fastest Apollo entry to splashdown time of 12 minutes 58 seconds, hitting the water at 35.4 kilometres per hour instead of the more normal 30.5 kilometres per hour.

Figure 6.34. Two hundred years apart, two very different ships called *Endeavour* sprang leaks on their voyages. To celebrate the occasion Honeysuckle Creek's Station Director Don Gray presented the Apollo 15 crew with a poster of an extract from Captain Cook's journal describing his incident, which occurred in the Great Barrier Reef, and a picture of Honeysuckle Creek taken by the author during the mission. Dave Scott (left) and Al Worden (right) accept their copies from Don Gray.

Figure 6.35. Approval from the House of Funds. Congress applauded the Apollo 15 crew with a standing ovation after mission Commander Dave Scott had finished his address to them on 9 September 1971.

The Apollo 15 crew took a small block of wood from the sternpost of Captain Cook's *Endeavour* and left behind a falcon's feather and a four leafed clover for Irwin's Irish ancestry. So ended what some of the top scientists felt was, "one of the most brilliant missions in space science ever flown."

At their first public appearance after their return, Dave Scott said, "I'd like to quote my favourite statement, which I think expresses our feelings since we've come back: 'The mind is not a vessel to be filled, but a fire to be lighted.'"

Scott, who spent more time in space during the Apollo missions than any of the 29 Apollo astronauts, later wrote, "Occasionally, while strolling on a crisp autumn night or driving a straight Texas road, I look up at the Moon riding bright and proud above the clouds. My eye picks out the largest circular splotch on the silvery surface, Mare Imbrium. There, at the eastern edge of that splotch, I once descended in a spaceship. Again I feel that I will probably never return, and the thought stirs a pang of nostalgia."[52]

1972 Apollo 16: a Routine Mission

Ken Mattingly overheard some youngsters say that the astronauts in their suits looked like Casper, the friendly ghost, so decided to call the Command Module *Casper* for a touch of humour and so that kids the could identify with the mission. Charlie Duke, at 36 the youngest of the astronauts to walk on the Moon, explains the Lunar Module's name, "We had considered names like sailing ships or explorers, but nothing really struck our fancy. We

Mission Data: Apollo 16

Date:	16–27 April 1972
Craft names:	*CASPER* (CSM) *ORION* (LM)
Craft numbers:	AS-511/CSM-113/LM-11
Personnel:	John Young Ken Mattingly Charles Duke
Duration:	11 days 1 hour 51 minutes 5 seconds
Features:	64 Lunar orbits in 5 days 5 hours 53 minutes 71 hours 2 minutes on Lunar surface
EVAs:	3, totalling 20 hours 40 minutes
EVA distance travelled:	26.7 km
Lunar samples:	96.6 kg
CSM weight:	30,368 kg
LM weight:	16,437 kg
Landing area:	8.99° S, 15.51° E in the Descartes Region
LM impact:	Area and time unknown
Distance travelled:	2,238,597 km

then decided we would like a constellation for a name, and *Orion* was short and easy to pronounce, so chose *Orion*."

"Probably one of the few constellations we knew about," added Young.

The mission insignia depicted an American eagle and a red, white and blue shield to pay tribute to the people of the United States. Crossing the shield was a gold NASA vector and sixteen stars to represent the mission number.

The Descartes site was almost 2,450 metres higher than the Apollo 11 site. "We kinda think of it as landing on the top of the Andes Mountains," John Young, commander of the mission, and the first man to go into lunar orbit twice, said before they left.

Figure 6.36. "Can I reach down to pick up a small moon rock?" Charlie Duke tried the flexibility of the International Latex Corporation Industries' A7L-B lunar surface spacesuit. Shown in the picture is the pressure garment assembly, worn under the white thermal and micrometeorite protective suit so familiar in the astronauts' pictures on the lunar surface. A cable restraint across the chest and back helped keep the shape of the suit and assisted shoulder movements. This type of suit was first worn in the Apollo 15 mission.

Figure 6.37. "Now, let's see... have we forgotten anything? Gosh, I hope it all works up there!" The Lunar Rover for Apollo 16 stripped of its equipment and neatly folded up for the flight to the Moon.

Figure 6.38. Family Affair on Earth. In a benign moment, Launch Pad Leader Guenter Wendt, better known as the "Führer of the Pad", on the right, shows the Apollo 16 spacecraft to Lunar Module Pilot Charlie Duke (left) and his sons Thomas (4), Charles (6), and wife Dorothy.

Following a routine launch from Pad 39A at 12:54 pm EST, Apollo 16 arrived at the Moon and went into orbit. Mattingly called Houston, "Henry (Hartsfield), it feels like we're clipping the tops of the trees."

Duke describes their impressions, "It did feel like we were right down in the valleys. I couldn't believe how close we were to the surface ... we were rocketing across the surface at about three thousand miles per hour in this low orbit, with mountains and valleys whizzing by. The mountain peaks and craters went by so fast, it gave you the same impression as looking out your car window at fence posts while traveling at seventy miles per hour."[54]

Young and Duke climbed into the Lunar Module to descend down to the lunar surface. Duke says, "I was having trouble with a leaky orange juice bag. NASA had designed a plastic drink bag to fit inside our spacesuits, since we were going to be working on the lunar surface for long periods at a time. It was shaped like a hot water bottle and attached to our long underwear. A long plastic straw went up from the bag, up through the neck ring of the helmet, right next to our mouths. To drink, we simply grabbed the straw between our teeth and sucked real hard.

"Well, right before our separation manoeuver, I had donned my helmet and immediately my drink bag began to leak through the straw into my helmet. It seemed like every time I breathed, out would come one or two small drops of orange juice.

"I couldn't suck them, I couldn't reach them with my tongue. I could only watch cross-eyed as they floated out in front of my face. Eventually some of the drops would hit me on the tip of the nose and slowly migrate up into my hair, giving me a sticky orange juice shampoo. It was really frustrating not being able to wipe the stuff off, as it touched and tickled my nose."[54]

Then just after separation, with *Casper* ahead of *Orion* as they went behind the Moon, *Casper* had to make a burn to change orbit.

Mattingly contacted Houston. "There is something wrong with the secondary control system in the engine. When I turn it on, it feels as though it is shaking the spacecraft to pieces."

This was serious – that engine was their ride home! Young thought hard and though he hated to say it, ordered, "Don't make the burn. We will delay that manoeuvre."

Their hearts sank down to their boots – two and a half years of training and only 12.9 kilometres from their target and now it looked like they would have to abort and return back to Earth. The two spacecraft circled the Moon in company, anxiously waiting for an answer from Houston.

Duke recalls, "We knew in our minds it was very grim. It looked as if we had two chances to land – slim and none. We were dejected."

"It was a cliff-hanger of a mission from where we were sittin' in the cockpit," Young said. "The secondary vector control system on the SPS motor wasn't workin' right and if

Figure 6.39. A view from the Moon on 20 April 1972. With engine trouble and the nearest workshop or mechanic over three days away, Mission Control at Houston was needed. If Texas was on the other side of the Earth, Madrid in Spain or Honeysuckle Creek in Australia would be tracking and the signals were sent around the world to keep the flight controllers in constant touch with the spacecraft for them to decide the fate of the mission. *Casper*, seen in the picture, and *Orion* anxiously kept each other company circling the Moon until Mission Control analysed the problems, and then advised them to go ahead and land.

Figure 6.40. "Yeow, this is some crater... it's really steep. I can't even see the bottom right where we are," called Charlie Duke as John Young headed for the edge of Plum Crater (on the rim of Flag Crater).

they didn't work right the mission rules said it was no go. The people on the ground did studies at MIT and Rockwell and in the end it worked out just fine."

There was no danger, said Houston, even if the engine backup control had to be used. "You are *go* for PDI."

Now six hours behind schedule, *Orion* headed down for the Cayley Plains. Spotting their shadow on the lunar surface as they came down to land, Young announced, "Boy, with that shadow, you really don't need the landing radar, and you can get a feel for the size of those craters that we were comin' over." Although *Orion* raised a lot of dust, it never obscured their view of the surface to the extent of the previous missions. Their arrival time was 8:23 pm on 20 April.

Duke remembers, "We were as excited as two little five year olds on Christmas morning. Imagine the best Christmas, the best birthday, the best visit to an amusement park – all rolled into one instant of time – that is the feeling we had as we

33

tried to describe what we were seeing." In contrast to Apollo 11's urgency to get out and walk, Young and Duke turned in for a sleep before setting foot on the lunar surface, mainly because of the long day coming up, and because of the six hour delay they had been up almost 20 hours. The delay also cost a lot of effort in changes to the excursion planning, and for a while threatened to scrub the third drive in the lunar rover to North Ray Crater because of the shortage of consumables.

Duke continues, "I began to write, and I wrote and wrote and wrote. It seemed like I was about to run out of pencil lead because of the hundreds of changes – changes in our time lines and changes in our procedures."[54]

After commenting to Capcom Tony England, "Fantastic. That's the first foot on the lunar surface. It's super, Tony," 36-year-old Duke joined Young to help pull the excursion equipment out. When they turned around they were startled to see they had landed *Orion* only 3 metres away from a 7.6 metres deep crater.

Young comments, "It would have been bad if we had landed in that crater. I saw it for a little while when comin' down, but where we landed it was perfectly flat, in the bottom of this 75-metre wide crater."

If they had landed on the rim of the deep crater they could have toppled over the edge, and that would mean they couldn't lift off – they would have been marooned on the Moon forever. They frequently talked about the close encounter with that crater.

Young: "I can't believe that big hole back there."

Duke: "John, you picked the exact bottom of this old crater."

Young: "There weren't any flat places around here, Charlie."

Duke: "Yes, but anywhere else we would have been on a great big slope."

The first excursion looking for Plum crater was frustrating, not being sure where they were.

Duke admits they were uncertain. ```There it is over there ... no, here ... no, no, that's not it,' – and we'd drive little further." They stopped at what they believed was Plum Crater, where they picked up some samples from the rim.

"Yeow ... this is some crater, Tony ... it's really steep. I can't even see the bottom right where we are," Duke called out.

"That is spectacular," Young agreed.

"Charlie, don't fall in that thing," warned England.

"I'm not gonna fall in," Duke said with feeling. Looking over the edge the two astronauts could see the steep sides and powdery dust could trap anyone trying to climb out, and with no ropes they realised there would be little chance of rescue if one of them did fall in.

Full of enthusiasm for the little lunar rover, Duke said as they returned to the Lunar Module, "You are making great time, John. We are doing 11 clicks (kilometres per hour)."

Houston: "Outstanding!"

Duke: "Super!"

Houston: "The Grand Prix driver is at it again."

Duke: "Barney Oldfield."

Young: "I could follow a road." – referring to the tracks made by the rover on its way out to Flag Crater.

Back at the Lunar Module after the first excursion, Young put the rover through its paces in front of the movie camera. Duke described the scene: "He's got about two wheels on the ground. It's a big rooster tail out of all four wheels and as he turns, he skids the back end, breaks loose just like on snow. Come on back, John ... I've never seen a driver like this. Hey, when he hits the craters it starts bouncing. That's when he gets his rooster tail. He makes sharp turns. Hey, that was a good stop. Those wheels just locked."

Young explains, "We drove it to see how it worked. We had to go up the side of a mountain with slopes more than 20°, and I think we did that because we bottomed out the pitch meter. We wanted to see how the vehicle handled. We had the camera there to document it too, which nobody else had done before. It was like driving on ice when you cut the thing too sharp at about 5 or 7 kilometres per hour, it would slide out and go backwards. The stuff on the Moon is very slippery. You don't hear anything but your suit pumps going when you're drivin' in a vacuum. It was very difficult to get in and out of – the Apollo 17 guys had a scoop to pick up rocks up without even stoppin' the rover."

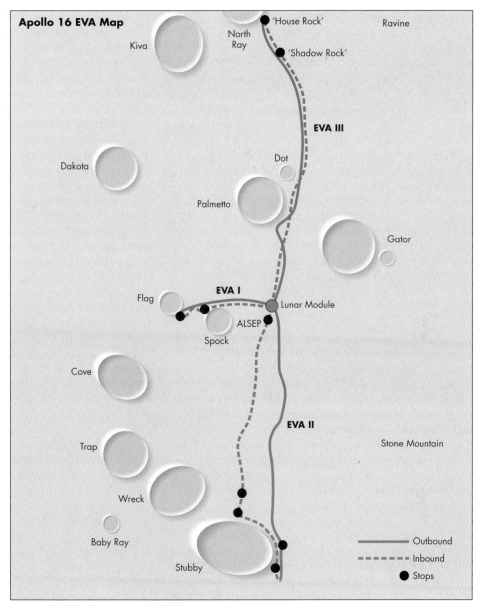

Figure 6.41. Route map of the Apollo 16 excursions.

Apollo 16 suffered a number of annoying problems. While Duke was setting up the ALSEP heat flow experiment, Young inadvertently yanked out a lead with his boot, which put an end to that experiment. For six years geophysicist Mark Langseth had been studying how much heat was being conducted out of the Moon's interior. His first experiment had burned up in Apollo 13's *Aquarius*; then Dave Scott wasn't able to drill down the required 3 metres in Apollo 15. Now Apollo 16 was a dead loss.

Both astronauts suffered sore fingers trying to work equipment and pick things up. Young thought it may have been because he had forgotten to trim his fingernails, a

Figure 6.42. "I've never seen a driver like this – hey, when he hits the craters it starts bouncing," called commentator Charlie Duke during the first Grand Prix on the Moon. Driver John Young found, " ...the stuff on the Moon is very slippery," when he gave the Lunar Rover a workout on the lunar surface during the third EVA, reaching a speed of 10 kph. Later, on a downhill run, it managed to reach a lunar speed record of 17 kph.

problem suffered by the Apollo 15 astronauts, but as Duke had trimmed his and he also suffered, it seems that it was a hazard of the pressurised space-suit.

Duke says, "Working in that spacesuit, squeezing those gloves and pressing the tips of our fingers against the ends, caused our fingers to seem like bloody stumps." The astronauts had to exert pressure all the time to keep the fingers bent when holding an object, or the suit pressure would pop the hand open.

Young and Duke both found difficulty drinking the water and orange juice from the suit containers. After 7 hours out on the lurain, Young says, "The first thing I wanted was a drink of water. I could have finished all my drink if I had a mouth behind my left ear. That was my only problem. It got lodged back there and I never could get at it."

In an attempt to overcome the potassium loss suffered by the Apollo 15 astronauts, Young and Duke were encouraged to take as much orange juice as they could, until Young finally confided with Duke with one of Apollo's classic passages: "I got the farts again. I got 'em again, Charlie. I don't know what the hell gives 'em to me. Certainly not ... I think it's acid in the stomach. I really do."

Duke: "It probably is."

Young: "I mean, I haven't eaten this much citrus fruit in twenty years. And I'll tell you one thing, in another twelve ... days, I ain't never eating any more oranges. And if they offer to serve me potassium with my breakfast, I'm going to throw up. I like an occasional orange, I really do. But I'll be damned if I'm going to be buried in oranges. ..."

Capcom: "*Orion*, Houston."

Young: "Yes, sir."

Capcom: "Okay, John. You're ... we're ... you have a hot mike."

Young: "H ... How long ... how long have we had that?"

Australia Talks to Apollo on the Moon

Then came an unscheduled event. Just before setting out on the second excursion the two explorers were discussing the plans for the day with Houston. Young remembers, "We were just sittin' there in the LM talkin' to Houston when Honeysuckle called back."

John Saxon, on the Operations Console at Honeysuckle, explains, "There was an earthquake in Los Angeles and we lost all the lines to Houston for a considerable period. I was madly trying to re-establish lines to Houston, when the astronauts called Houston, and I had to respond. We had a chat for about five minutes; I guess I am the only person in the Southern Hemisphere that actually got to speak to anyone on the lunar surface. The conversation was mainly about beer."

Figure 6.43. Honeysuckle Creek Operations at work. From the left: Deputy Director Ian Grant, Operations Supervisor John Saxon and Deputy Director Mike Dinn providing the human interface between the station and Mission Control at Houston.

Saxon: "*Orion*, this is Honeysuckle. We have a comms outage with Houston at this time. Stand by one, please."

Young: "Okay, Honeysuckle – nice to talk to you. How are ya'll all doin' down there?"

Saxon: "We are doing great. Nice to talk to you. We will be with you shortly, we are just getting some lines configured for you."

Young: "Have a Swan for us."

Saxon: "Say, again, *Orion*. You are pretty poor quality on this back up."

Duke: "Honeysuckle, what John was saying was have a Swan for us."

Saxon: "Oh, Roger. We should get the lines restored very shortly for you. Sorry about the delay."

Young: "Okay – you guys are nice to talk to. We do not care about the delay."

Saxon: "Thanks very much. Certainly appreciate it. It's a pleasure working on this mission."

Young: "Roger. We would like to come down there and see you folks at Honeysuckle."

Saxon: "Right – you've got a permanent invite, anytime you like."

Young: "That's very kind."

Saxon: "We will keep the beer cool for you."

Duke : "You have just got a couple of fellows ready to show up on your lawn. That's the best idea I've heard for a long time."

Saxon: "I think that's a pretty good one down here too."

Duke : "You see in my terminology that's certainly 48 packs. Right now that's how I feel – really love one."

Saxon continues, "This was picked up by the American press, and they published a few things about astronauts talking to Australia and talking about Swan Lager. Swan got hold of this, and thought it was marvellous publicity, so they sent each of the astronauts a crate of beer. We rang them up and said, 'Hey – look we're having a party too,' so we ended up with 48 crates of Swan beer!"

During the second excursion they climbed up a 20 degree slope to the Cinco Craters on Stone Mountain and reached a height of 152 metres above the valley floor, the highest level of any of the Apollo excursions. Looking back over the steep ridges below they found the panoramic view across the valley stunning.

"We can see the Lunar Module … look at that, John," an excited Duke shouted as he spotted the colourful *Orion* sitting in the middle of the drab gray valley floor about 5 kilometres away. The brilliant white of South Ray Crater dominated the view, with black and white boulders scattered across the landscape. Young was keen to visit it, but Houston figured the hazards were too great.

As the third excursion began, Capcom Tony England cheerfully announced, "Out again on that sunny Descartes Plains."

Young retorted, "Ain't any plains around here, Tony. I told you that yesterday."

At North Ray Crater they found a large rock they called House Rock which they estimated to be 27 metres long and 13.7 metres high. Unable to break a sample off it, Duke broke samples off a rock next to it someone called the Outhouse Rock. On the way back they set a Moon speed record of 17.1 kilometres per hour as their speedometer went off

Figure 6.44. Tourist Charlie Duke snaps a picture of a twentieth-century campsite on the Moon. John Young jump salutes the American flag while the rover rests beside the astronauts' home, the Lunar Module *Orion*. Stone Mountain rises in the background.

scale high. Duke felt they were going to launch themselves into orbit.

Before leaving to return, Duke decided, "… to take this time to place our family picture on the Moon, so I walked about thirty feet from the LM and gently laid our autographed picture of the Duke family on the gray dust. As I made a photograph of it lying there, I wondered, 'Who will find this picture in the years to come?'" He then added a special Air Force medallion to commemorate their silver anniversary.[54]

Figure 6.45. Family affair on the Moon. Duke said, "… I walked about thirty feet from the LM and gently laid our autographed picture of the Duke family on the gray dust. As I made a photograph of it lying there, I wondered 'who will find this picture in the years to come?'"

Duke described the moment of lift off at 7:25 pm on 23 April: "The Lunar Module is held together by three large bolts and at lift off the bolts explode, separating the ascent stage from the descent stage. At the same time a guillotine-like affair cuts the electrical, oxygen, and water lines.

"When the bolts exploded, instead of being propelled upwards, we dropped. 'Oh, no, it *didn't light and we are dropping!*' flashed through my mind.

Then – bang! – the engine ignited and instantly there was 1,588 kilograms of thrust. A kick at the bottom of my feet, and off we went – straight up for 244 metres!"[54]

After meeting up with Mattingly in *Casper* and on their way back to Earth, a large film canister of pictures from lunar orbit had to be retrieved from the SIM Bay at the back of the Service Module. After depressurising and opening the hatch, Mattingly climbed out.

Duke followed, and describes the scene for us: "I followed, floating out a body length, and anchored my feet on the hatch sill. My job was to make sure his safety line, plus oxygen and communication lines didn't get tangled in parts of the spacecraft.

"As I floated out, I was again overcome with the awesome beauty of space. The panorama of the universe was spread out before me, and I felt like a spectator in an audience watching the play unfold. Ken was the performer – and the universe was the stage.

To the right was the Earth, 318,641 kilometres away. It was a crescent Earth – just a thin sliver of blue and white – yet breathtaking to behold. Over my left shoulder was the Moon, only 67,590 kilometres away and enormous. It was a full Moon, and I could see clearly all the major features – the Sea of Tranquillity where Neil and Buzz had landed, Ocean of Storms, even the Descartes highlands. It was spectacular!

"Everywhere I looked it was blackness – the empty blackness of space, so powerful it seemed I could reach out and touch it. The feeling of detachment I experienced was strange; it was almost euphoric, and I wondered what it would be like to float off into this blackness."[54]

Mattingly was experiencing similar feelings as he collected the film from the SIM Bay. He was very aware of the vast nothingness surrounding the life supporting infinitesimal speck they had called *Casper*. Tightly gripping the handrail he could feel the comforting security of the spacecraft flow through his gloves and fingers. As they cruised back to Earth, he remarked: "There's not a scene on the Moon that carries the emotional impact of watching your Earth shrink to a little ball."

Watched by television, *Casper* dropped into the Pacific at 1:45 pm Houston time on 27 April, 2,414 kilometres south of Hawaii and flopped upside down before righting itself, 5 kilometres from the USS *Ticonderoga*.

Until the Apollo 16 mission the geologists were able to predict the type of soil the astronauts would bring back. The Descartes samples ended this run. Confidently predicting soil and rocks with a volcanic origin, the geologists were taken aback to find the samples turned out to be impact breccias. As Geologist Don Wilhelms admitted, "… we goofed."

On 17 July 1972 only three months after it was set up, Apollo 16's seismometer registered the largest impact ever recorded on the Moon when a meteorite hit the far side of the Moon near Mare Moscoviense.

1972 **Apollo 17 Takes the Last Astronauts to the Moon**

Apollo 17 was the last Apollo mission to the Moon, so the spacecraft were named with appropriate dignity, the Command Module *America* as a tribute to the mission and the American public. The Lunar Module was called *Challenger* because of what the future held for America.

The patch for Apollo 17, drawn by artist Robert McCall, was full of symbolism. Cernan explains the significance of the design, "We felt certainly that Apollo 17, in spite of the fact that it's the last flight in the Apollo Program, it's really not the end, but rather the beginning. It's

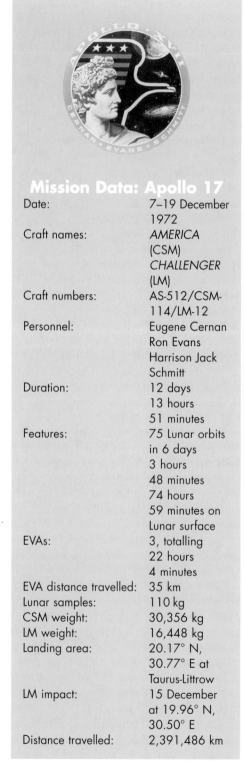

Mission Data: Apollo 17

Date:	7–19 December 1972
Craft names:	*AMERICA* (CSM) *CHALLENGER* (LM)
Craft numbers:	AS-512/CSM-114/LM-12
Personnel:	Eugene Cernan Ron Evans Harrison Jack Schmitt
Duration:	12 days 13 hours 51 minutes
Features:	75 Lunar orbits in 6 days 3 hours 48 minutes 74 hours 59 minutes on Lunar surface
EVAs:	3, totalling 22 hours 4 minutes
EVA distance travelled:	35 km
Lunar samples:	110 kg
CSM weight:	30,356 kg
LM weight:	16,448 kg
Landing area:	20.17° N, 30.77° E at Taurus-Littrow
LM impact:	15 December at 19.96° N, 30.50° E
Distance travelled:	2,391,486 km

sort of a conclusion of the culmination of what we consider man's greatest achievement, certainly in our lifetime. And, looking into the future, these achievements and the potential of them have literally no bounds. So, we have a bust of the god Apollo on our patch. He represents not just Apollo and the Apollo program, but we feel that he represents mankind himself.

"He represents knowledge and wisdom; Apollo is looking out into the future. He is not looking behind. And he's not simply looking at the Moon – someplace that mankind has been, and in a sense a goal that mankind has accomplished, but he is looking beyond the Moon and into the future.

"We have along with him, up in the corner of our patch, a golden Moon, sort of representing a golden era of spaceflight that we are bringing to a close now.

"Superimposed upon this Moon, alongside the bust of Apollo, alongside mankind, we're a little bit parochial: we have a very contemporary American eagle whose wings are coloured with blue and red stripes of our flag. The eagle's wing just touches the Moon to suggest it has been visited by man, and we have three white stars representing the crew indented into the top of the eagle's wings.

"The significance there is to remind us – not just in this country, and as I say, parochially speaking – the rest of the world, that the achievements that have happened in the past decade were not by accident. America brought us to where we are today and the United States of America is going to lead us into the achievements and accomplishments of the future."

The crew also included a politically chosen member when Joe Engle was pushed aside for scientist/geologist Harrison Schmitt by the bureaucrats in Washington. Despite vehement protests by Slayton and Cernan, political power prevailed and Jack Schmitt joined the crew as Lunar Module pilot. Although not a test pilot, he did qualify as a pilot of the T-38 trainers, but he was a recognised authority on photographic and telescopic mapping of the Moon. He was also an instructor on lunar geology to the astronauts. After some initial confrontations and rebelliousness Schmitt settled down to become an excellent and respected member of the Apollo 17 team. With little in common, test pilot Cernan was often puzzled by geologist Schmitt's reaction to events and comments but agreed that Dr Rock, as he called him, overcame some major difficulties in the alien environment of the aviation industry, admitting that his encyclopaedic knowledge as a professional geologist on the lunar surface was an asset to the mission's science results.[58]

Where should the last lunar mission of the Apollo era land to gain the maximum benefit for science? The primary objectives were to gather old highland material; young volcanic material; and obtain new observations from orbit not covered by previous missions. Landings near the Tycho and large Copernicus craters were always attractive, but were rejected; Tycho because of the hazards of the rough terrain, and Copernicus because three landings had already occurred in the vicinity of Mare Imbrium, and some Apollo 12 samples may have provided an age for this impact event. A daring landing on the far side in the Tsiolkovsky crater was considered but the cost and time of providing a communications satellite made that choice unfeasible. Finally three sites rose to the top of the list – Alphonsus Crater, Gassendi Crater, and the Taurus-Littrow Valley. Although Gassendi was the most scientifically interesting site, there were no known regions of young volcanism near the landing site, so at the time of the Apollo 16 mission Alphonsus was the scheduled site.

Taurus-Littrow eventually won the day because material from a landslide along the southern wall of the valley was expected to provide the best desired samples of ancient highland material as well as allow determination of the age of the Serenitatis impact. Samples from the floor of the valley would allow the age of the mare material to be

determined. From orbital observations, particularly by Al Worden during Apollo 15, it seemed several craters in the region could to be surrounded by relatively young volcanic cinder cones, one of them the crater Shorty. A bonus was if the Lunar Rover failed the primary objectives of the mission could still be met.

A normal Apollo mission plan would have ended in landing the astronauts and their spacecraft on the Moon during a solar eclipse, putting the spacecraft in shadow for up to nine hours. The planning engineers felt that some of the spacecraft systems might not survive such a long cold period, so to arrive at the target in sunlight with the Sun at the right angle, Apollo 17 was scheduled to have the first night launch.

Scheduled for a 9:53 pm lift-off, Apollo 17 had the only last minute hold of the Apollo launches. As the last moments approached the astronauts steeled themselves for the thrill and excitement of lift-off and heard the count drop to "Thirty …" and stop. The count had stopped at thirty seconds to go! Cernan's fingers tightened on the abort lever – just in case. In the firing room a red light flared indicating the pressurization for one of the propellants in the Saturn-IVB hadn't registered because a ground computer failed to send a command to the third stage oxygen tank due to a faulty diode. When the manual override also failed the launch team began a frantic procedure to bypass the fault before time ran out. Countless prayers were answered when the count resumed within the launch window.

Figure 6.47. The last Saturn V bound for the Moon had just been sent on its way and the personnel in Firing Room 1 at the Kennedy Space Center stood to listen to an address by Vice President Spiro Agnew. It took 463 technicians in the firing room, backed by 5,000 consultants on standby, just to get the rocket off the ground and into orbit.

◀ **Figure 6.46** ."Whoopee!" cried Ron Evans as Apollo 17 rose from the pad – the last manned lunar launch of the Twentieth Century. Almost as bright as the Sun, the intense glow from the wake of the only Apollo night launch was seen from 800 kilometres away as the rocket gobbled over 13 tonnes of fuel per second. The crew in the Command Module only saw a lurid red glow bouncing off the clouds.

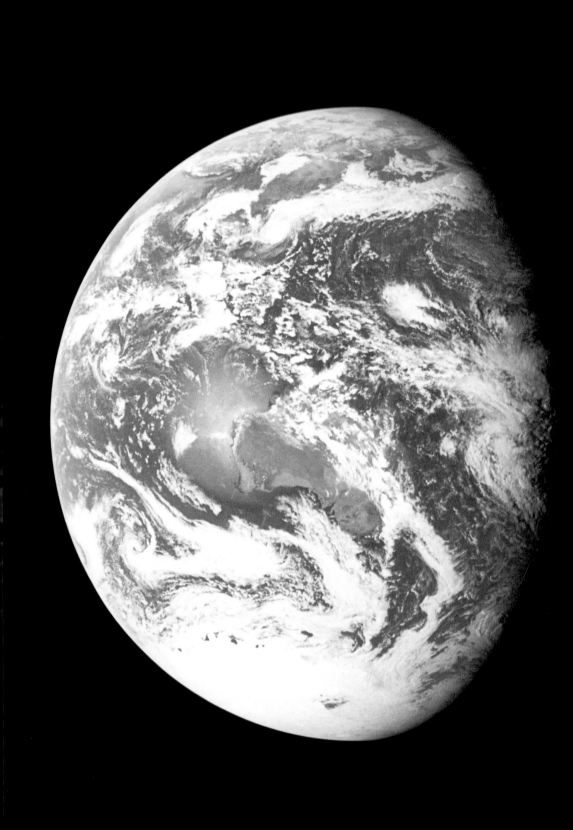

At last, Launch Control called, "Two … one … zero … we have a lift-off. We have a lift-off and it's lighting the area, it's just like daylight here at the Kennedy Space Center as the Saturn V is moving off the pad. It has now cleared the tower."

It was 12:33 am EST.

It was a brilliant spectacle. With an intensity equal to the Sun, the dazzling glare from the streaming 800-metre long wake of the giant rocket lit up the night sky and was seen as far away as Cuba and North Carolina, over 800 kilometres away. Inside the Command Module the astronauts were unable to see the big picture of the whole landscape around them lit up by their fiery tail but saw their instrument panels and suits glowing a lurid red from the rocket's glare bouncing off the clouds and shining through the Commander's window. An elated Evans yelled, "Whoopee!" and the normally reserved Schmitt called out, "We're going up – man, oh man!"

Apollo 17 was the first spacecraft to break out of Earth orbit and head for the Moon from over the Atlantic instead of the Pacific because the Saturn IVB rocket was not quite powerful enough under the conditions to push the spacecraft to escape velocity from the Pacific.

Apollo 17 followed the now familiar eighty six hour routine of travelling to the Moon and looping into lunar orbit, before separating from the Command and Service Module. Coming in to land Schmitt called the numbers, while Cernan took over control and skilfully steered *Challenger* into the Valley of Taurus-Littrow, a gorge deeper than the Grand Canyon. Slipping smoothly between the North Massif on their right and the South Massif on the left, the Lunar Module dropped to the valley floor, over an unexpected huge boulder and deep hole heading towards the crater Camelot. Beyond were jagged, tooth-like rocks rearing out of the ground. Cernan manoeuvred the Lunar Module onto a clear patch bet-

Figure 6.49. While they were looking at Australia, a brand new 64-metre dish antenna at Tidbinbilla, DSS 43, was tracking them on its first major assignment.

◄ **Figure 6.48.** On the way out to the Moon, the Apollo 17 astronauts looked back to see Australia emerging from under a cloud bank and took this picture with a 70 mm Hasselblad and a 250 mm lens. The specular reflection from the Sun is visible in the Indian Ocean off the coast of Western Australia.

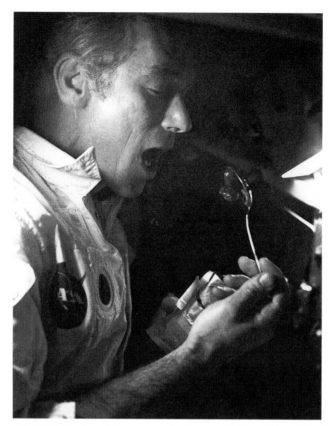

Figure 6.50. Weightless eating: by Apollo 17 Gene Cernan had become adept at manipulating his food in zero-gravity.

ween automobile sized boulders and landed onto the charcoal-gray surface with a stomach-jolting thump just 61 metres from the planned landing point at 1:54 pm on 11 December 1972.

"Gene landed the LM as if it were an everyday event," applauded Schmitt. Four hours later, coming to the end of his EVA checklist, he then announced, "The next thing it says is that Gene gets out!"

Cernan asked Schmitt, "How are my legs? Am I getting out?"

Schmitt replied, "Well, I don't know. I can't see your legs … I think you're getting out though, because there isn't as much of you in here as there used to be."

Cernan felt a great satisfaction and sense of achievement to be able to plant a Cernan bootprint on the lunar surface and looking around at the looming mountains, giant boulders, landslides and craters found to his pleasure they had landed beside the crater he had named after his daughter and reported, "I think I may just be in front of Punk."

Cernan noticed the soil glittered with what looked like millions of tiny diamonds, but the magic evaporated when Schmitt joined him and reported he was seeing specks of glass. "The soil looks like a vesicular, very light-coloured porphyry of some kind; it's about ten or fifteen percent vesicles."

The now familiar routine of exploring around the Lunar Module in the rover was interrupted by Cernan breaking part of the wheel fender off with his rock hammer sticking out of the pocket of his suit. "Yeah, I caught it under my hammer. The reason it was so important to fix it was because of the lunar dust. It's fine like graphite, but rather than a lubricant, it's a friction producing material – it gets into everything, into your visor, into the electronic gear, and when we drove the rover without that portion of that fender we had a rooster-tail of dust thrown completely over the top – over everything, and that was just unacceptable. So we made a fender out of some geology maps. We took duct tape, but we couldn't use it because of all that lunar dust, we couldn't clean it off enough for the tape to stick. So we taped a couple of maps together the night before and then had to use light clamps from inside the LM to clamp it on to the existing portion of the fender. When we came home we needed the clamps because they held both lights, so we brought

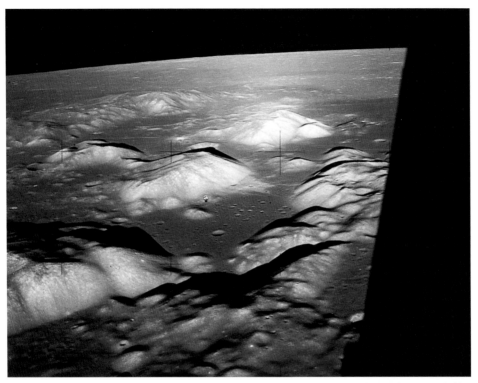

Figure 6.51. A spectacular view of the Valley of Taurus-Littrow and its ring of massifs surrounded by the plains of the Serenitatis Basin. The Sculptured Hills, the bumpy terrain in the foreground, are under the minuscule CSM *America* as it continued on to orbit the Moon, leaving the Lunar Module *Challenger* to find its own way down to the floor of the valley.

Figure 6.52. "Hallelujah, Houston. *Challenger*'s baby is on a roll!" After landing on the valley floor, seen under the Command and Service Module in Figure 6.51, Gene Cernan checked the stripped Lunar Rover was running before he and Schmitt assembled all the additional equipment required for their excursions. The slopes of South Massif rise on the right.

the fender home and it's now in the Smithsonian in Washington."

This was the first successful automotive repair on the Moon.

As Schmitt clambered down the ladder for their second excursion, he looked up and announced, "My, what a nice day. There's not a cloud in the sky!"

On their way back from the South Massif near Shorty Crater on this trip, Schmitt was about to take a photograph when his excited voice stirred everybody up. "Hey – it's orange. I found orange soil!"

Cernan thought his companion was running low on oxygen or he had been on the Moon too long, so called out, "Well, don't move till I see it."

Schmitt: "It's all over orange."

Cernan: "Don't move till I see it."

Schmitt: "I stirred it up with my feet!"

Cernan: "Hey – it is! I can see it from here."

Schmitt: "It's orange!"

Cernan: "Wait a minute, let me put my visor up … It's still orange!"

Schmitt's boot had kicked the ground and revealed soil ranging from bright orange to ruby red, which at the time was hoped to be more recent volcanic activity but turned out to be microscopic glass beads, tinted by titanium, about the same age as the rest of the rocks around. It had been ejected by an impact, not by volcanism.

At 13° Apollo 17 had the highest Sun angle of all the missions. Cernan says, "When you are on the surface of the Moon in the daytime it's a paradox. You are standing on the surface of the Moon lit by sunlight – you, your body and the surroundings, and you look up at the sky and it's black – it's not darkness – it's just black. Most people confuse darkness with blackness – they are two totally different words. Darkness is the absence of light in my definition. Blackness is a void. Blackness is the absence of almost anything. If you look at the Earth from the Moon it reflects sunlight, yet it is surrounded by the blackest black you could ever conceive in your mind – the absence of anything. The blackness has three dimensions. I didn't find the black sky above oppressive. I define blackness as the infinity of time and space and if you let your mind and imagination wander the infinity of time and space does anything but close in upon you. When you stand on the Moon and look up and see that blackness which goes all the way to the horizon of the Moon, it doesn't feel like you are being closed in upon like a black painted ceiling at all – as a matter of fact it is exactly the opposite – you know it goes on forever.

"When you are on the Moon you can't look anywhere near the Sun – it's devastatingly bright. When we drove the rover back to the east it was a lot more difficult to see up-sun than down-sun because of the reflective surface. The closer you looked toward the Sun you just couldn't see much definition at all.

"A lot of people say can you see anything else in the daytime on the Moon – can you see stars? The answer to that is yes – if you shield your face and eyes from all the reflected light around you can see stars in the daytime on the Moon – not as brightly as at night of course."

A visit to the North Massif during the third geological excursion during day three and a visit to the Sculptured Hills and the Van Serg Crater brought to an end the last journey on the surface of the Moon in the twentieth century. By this time both Cernan and Schmitt were weary, aching, and rubbed raw trying to follow all the planned instructions and changes relayed up from the geological experts gathered at Mission Control in Houston.

Houston called to the moonwalkers, "Okay, you guys, say farewell to the Moon."

Cernan replied, "Bob, this is Gene. I'm on the surface … as we leave the Moon at Taurus-Littrow, we leave as we came, and, God willing, we shall return, with peace and hope for all mankind."

Gene Cernan turned to climb the ladder and spotted a plaque mounted there by a Grumman factory worker and repeated the inscription aloud, "Godspeed the crew of

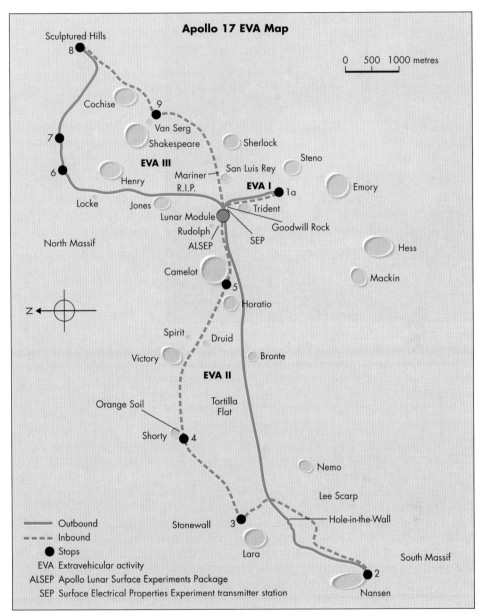

Figure 6.53. Route map for the Apollo 17 excursions.

Apollo 17." He then climbed up the nine steps of the Lunar Module's ladder to become the last person in the Apollo Program to leave the lunar surface. At the top he paused and looked around. "I felt excited that we had been there, but disappointed that we had to leave. Jack Schmitt and I described that valley that we landed in as our own private little Camelot. We knew once we left we would never come back. It was our home – it was a uniquely historical place no man had ever been before in the history of life on this planet of ours. You were there – you made your imprint. You would think that

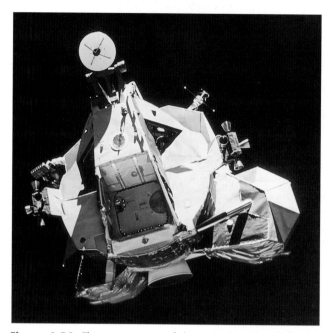

Figure 6.54. The ascent stage of the Lunar Module *Challenger* approaching *America*. Part of Gene Cernan's face is visible in the triangular docking window as he concentrated on bringing the two spacecraft together. The two returning astronauts' lives depended on a successful docking.

would be enough, but there was so much to do. Then you *do* leave and you remember all the things you wished you would have done – little things or big things – or whatever. It was hard to leave but it was time to leave. I always thought that if I knew things were going to go so well I wish I could have stayed another week or two. But you do know the longer you stay the more vulnerable you might become to problems that might come to keep you from getting home."

On Earth Mission Control read a statement from President Nixon: "As Challenger leaves the surface of the Moon we are conscious not of what we leave behind, but of what lies before us." So, as the last words exchanged between the Moon and Earth echoed around the world, what were the people of Planet Earth who were listening thinking?

It seems everybody remembers the first step on the Moon, and of course that is what the people on Earth commemorate, but few can remember the last person to pull his boot off the surface of the Moon in the twentieth century.

At 4:54 pm on 14 December the last unofficial words spoken on the Moon's surface were heard: "Okay, Jack, let's get this muther outta here," as Cernan flicked the yellow ignition switch and red flames ripped into the lunar surface. Shredded gold foil from the descent stage glinted in the boiling cloud of gray dust shooting out from under the engine bell housing. The Stars and Stripes whipped madly in the rocket's exhaust, then relapsed into a permanent stillness as the rocket's red glare dwindled into the distance above, and winked out. The dust drifted down to settle over the discarded twentieth century artefacts. The last of the aliens had gone.

Apollo 17 returned to Earth to splash down in the Pacific at 1:24 pm spacecraft time on 19 December. The crew of Apollo 17 were welcomed back with a big party on the carrier USS *Ticonderoga*, and entered the record books with the longest manned flight to the Moon, the heaviest swag of lunar samples; the longest activity time on the lunar surface with the greatest distance travelled, the longest time in lunar orbit, the greatest distance travelled and the only Saturn V night launch.

Figure 6.55. The last lunar mission splashdown: a bird's eye view of Apollo 17 smacking into the Pacific Ocean.

Back on the Moon a timeless stillness descended on the floor of the valley of Taurus-Littro. As with the other Apollo landing sites, there had never been any sounds of the visit of the twelve humans because there is no air on the Moon. Only the Sun will rise each day to reveal the six deserted campsites with their human bootprints and tyre tracks, frozen motionless to remain just as they were left. Will they remain undisturbed until the next visit by people from Earth ... perhaps by another form of life ... or perhaps abandoned forever?

The answer lies somewhere in the future.

Figure 6.56. Winding up the Apollo lunar missions. The beginning and end of each mission travelled at a snail's pace. The Apollo 17 crew go from 38,600 kph during re-entry to creeping up to the recovery helicopter at winch speed as they were brought back to civilization. The USS *Ticonderoga* stood by in the background to fete the last men from the Moon.

Figure 7.17. At this point Honeysuckle Creek left the Manned Space Flight Network for the Deep Space Network and supported the Voyager encounters with Jupiter and Saturn, the Viking Mars landers, and the Venus probes, as well as other projects.

8 Apollo–Soyuz – The End of the Apollo Era

While the abandoned Skylab was still spinning around the Earth, accompanied in space by a Russian manned Salyut space station, the Americans and Russians finally shook hands in the doors of their spacecraft over Verdun in France, for a moment dropping the barrier between Communism and Democracy with the politically motivated Apollo–Soyuz mission. What ever happens in the future, this act tied up the rivalry of these two great space-faring nations and neatly sealed the first era in space.

The ASTP mission was originally conceived in the Nixon–Kissinger days, and agreed to by President Nixon, Soviet leader Leonid Brezhnev and A. Kosygin, Chairman of the Soviet Council of Ministers, on 24 May 1972. The mission became a focal point for a number of factions. The Russians saw an opportunity to observe the mainstream of US space technology, and to appear to the world to be at least on equal terms with the Americans by flying a joint mission. They also saw the possibilities of developing a technical base for an international space

Figure 8.2. The spirit of Apollo–Soyuz. Slayton and Leonov show weightless comradeship between the astronauts and cosmonauts. One might speculate: what would Premier Khrushchev have thought of this mission?

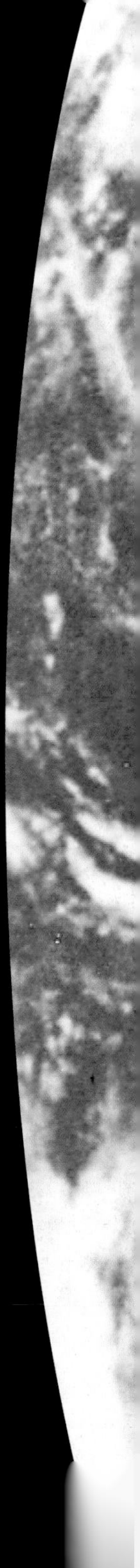

Figure 8.1. Soviet blastoff: on the morning of 15 July 1975, the three stage rocket with the Soyuz 19 spacecraft headed for space and the historic first international rendezvous. It was also the first time the

rescue system. The Americans objectives seemed to be mainly political, but they recognised the usefulness of rescue in space, and it was another opportunity to encourage scientific support for manned space flight.

The Apollo–Soyuz mission was simply an experiment. There was no new technology involved – both countries were using old spacecraft and conventional operational techniques. The Apollo spacecraft was the earlier "H" series model, not the later "J" series. The Soyuz (Russian for Union) had been developed about the same time as Apollo. Formidable technical, linguistic, and operational challenges had to be overcome before the two spacecraft could get off the ground. The open American philosophy clashed against the covert Russian outlook. Tom Stafford refused to fly the mission unless he and his crew were allowed to inspect the Russian Soyuz spacecraft, which was eventually approved.

Apollo–Soyuz was a strange mission for the Americans – a mixture of speeches,

Apollo–Soyuz

Apollo

Date:	15–24 July 1975
Personnel:	Tom Stafford
	Donald "Deke"
	Slayton
	Vance Brand
Duration:	9 days 1 hour
	28 minutes
Features:	138 orbits
Apogee:	230 km
Perigee:	219 km
Spacecraft weight:	6,739 kg

Soyuz

Date:	15–21 July 1975
Personnel:	Aleksei Leonov
	Valery Kubasov
Duration:	5 days 22 hours
	30 minutes
Period:	88 minutes
Apogee:	225 km
Perigee:	222 km at
	rendezvous

hand shaking, gift exchanging, television covered politics on one hand and hard working, laboratory experimental science on the other. The Russians insisted the link-up should be over Russia, while the Americans wanted it over America, and it ended up over France and West Germany, where they could both see their respective tracking stations.

This mission was also the last opportunity for Deke Slayton to fulfill his dream of flying in space. Ironically, although one of the seven original Mercury astronauts and now the Senior NASA Astronaut, he became the oldest space rookie at the age of 51. After a constant battle to get back onto flight status with the doctors and management over a slight disturbance of the regular rhythm in the upper chambers of his heart, Slayton found that taking lots of vitamin tablets to beat a cold seemed to stop his fibrillation's.

After a solid week of tests with heart specialist Dr Hal Mankin he was cleared and on 13 March 1972 returned to flight status. With his age, the last Apollo crews already in training, and the Skylab crews announced, his options for a flight boiled down to the Apollo–Soyuz flight. So when Chris Kraft, Director of the Manned Spacecraft Center, asked Slayton for his crew recommendations for the Apollo–Soyuz flight, he submitted his own name as commander, with Jack Swigert and Vance Brand.

However, when he was called to Kraft's office just after Apollo 17, he found Tom Stafford replaced Jack Swigert and was appointed the Commander because of his space flight and rendezvous experience, and he already had a good relationship with the Russian cosmonauts, as well as their trust.[29]

Stafford says, "I lived in Russia for six periods at a time. Minimum for about two weeks, maximum about six weeks. The Cosmonauts came to the United States for five times for about a month each. We trained at Star City near Moscow in the simulators and procedures trainers on their systems, the language … the whole thing."

Orroral Valley Prepares

Orroral Valley, over a mountain ridge from Honeysuckle Creek, was the main Australian support for the Apollo–Soyuz mission, as Honeysuckle Creek was already an established Deep Space Station, and Carnarvon had gone. As Orroral Valley had no voice sending or receiving equipment, new equipment had to be installed.

Ian Fraser, the Apollo–Soyuz Coordinator in Australia, explains, "During the Apollo and Skylab missions, Orroral Valley provided support for the scientific spacecraft used to monitor the Sun's potentially lethal radiations, as well as weather satellites. When it was our turn to fully support manned flight, Orroral Valley's approach to Apollo–Soyuz was very much an outcome of its STADAN beginnings. Being a multi-link, continuous operation station supporting up to 40 satellite passes per day, to us, Apollo–Soyuz was just another spacecraft – except it was manned.

"For the Apollo–Soyuz we dedicated a team of operational specialists combined with selected maintenance personnel. Honeysuckle Creek was a 'wing' site with VHF voice support only, providing redundancy in the case of a USB system failure at Orroral Valley. We expected the Honeysuckle gear to be shifted across the mountain – but it didn't happen that way. It came from various places, and they extended the building to take the extra equipment."

Ian Edgar, the Communications Technician (Comtec), recalls, "For Apollo–Soyuz, the station ran the normal four shifts for the scientific satellites, normally about 28 people. There were only four of us operating the equipment for Apollo–Soyuz. We covered 12 to 14 hours with the one shift, we had camp beds set up in the training room, and for that whole period we didn't go home. The longest pass was about five minutes – most of them were only about two to two and a half minutes, some were only thirty seconds."

Figure 8.3. The Apollo–Soyuz American crew. From the left: Deke Slayton, Docking Module Pilot; Vance Brand, Command Module Pilot; and Tom Stafford, Mission Commander.

Two ARIA (Apollo Range Instrumentation Aircraft), modified Boeing 707 jet aircraft with 16 crew members in each, arrived at Perth Airport to cover the first three orbits of the Apollo spacecraft, much as they had done for all the previous Apollo flights. They left at 2:20 am and flew at a height of 9,144 metres for 7 hours, one aircraft 2,600 kilometres in a southwesterly direction to take over from an ARIA from Capetown, and one in a south easterly direction.

The Soviet Mission Control Centre was

located at Kaliningrad, about 45 minutes northeast by car from Moscow. Kaliningrad takes over from launch control at Baikonur after the spacecraft has separated from the launch vehicle in orbit.

The Apollo–Soyuz Mission

1975: 15 July Two Launches on One Day

Soviet television devoted a whole five hours of their programming to the mission, from Yuri Gagarin's flight to Apollo–Soyuz launch, the first live broadcast of a Russian lift-off. For days newspaper headlines told the people of the Soviet Union about the great spirit of co-operation between the two superpowers.

The day 15 July 1975 arrived with clear skies and light breezes promising a perfect launch for the Russians. Their Soyuz 19 spacecraft left the ground at 3:20:10 pm Moscow time from the same pad that launched the Sputnik and Gagarin missions. Afternoon in Baikonur was early morning in Florida where the Americans listened to the Russian launch and realised it was "Our turn to hit the ball. Now we've got to get into orbit." The American astronauts were still asleep while on the launch pad the Apollo launch rocket was gorging on liquid oxygen at a rate of 4,540 litres per minute. Although thunderstorms were hovering around Cape Canaveral they were not expected to affect the launch.

Deke Slayton, the man who waited for so long to get into space, who never gave up, was finally approaching his spacecraft. Thirteen years late, he was feeling good, though he had never meant to be the world's oldest astronaut.[29]

Seven and a half hours after the Soyuz launch the last Apollo spacecraft with three astronauts and a stowaway mosquito was launched from Cape Canaveral at 3:50 pm EDT, 6,224 kilometres behind the Russians, flying over Belgrade at that moment.

Two thousand skilled workers, with a perfect record of successful Saturn launches behind them, stood and watched "the Stack" and their jobs go. The seven-year-old rocket, the thirty-second and last Apollo to be launched, was the best booster as far as technical problems went. "We saved the best till last," said Chief Saturn test supervisor, William Schick.

Walter Kapryan, Director of Launch Operations lamented, "We are losing a team that I don't think we'll ever have together again." The Saturn launch facilities were maintained for another 12 months in case NASA reversed its decision to end the Apollo missions.

As they shot into orbit an elated Slayton called out, "Man, I tell you, this is worth waiting 13 years for! This is a helluva lot of fun – I've never felt so free."

Out in space aboard the Salyut 4 space station

Figure 8.4. Vladimir Shatalov, in charge of Cosmonaut Training, on the left, with Aleksei Leonov, Commander of the Soviet mission, and flight engineer Valeriy Kubasov, during training in the Mission Control Center at Houston.

above the Apollo–Soyuz mission Pyotr Klimuk and Vitaly Sevastyanov were woken up from a sleep to listen to the launch. Launched on 24 May, they were well settled into a 60-day mission. Later Sevastyanov called the Soyuz crew on their radio link through the Russian tracking stations for a long talk which included: "I think that all those who are in space right now are aware that this is a grandiose task, and there are seven people in space right now … the magnificent seven!"

First International Rendezvous in Space

By 8:00 am, 17 July the two spacecraft were approaching each other. When they were 322 kilometres apart Brand called Houston, "Okay, we got Soyuz in the sextant."

"Got a good view of him, Vance?" asked Capcom Richard Truly.

"He's just a speck now."

The Americans did all the manoeuvring as only they had enough fuel. It took 388 kilograms to rendezvous; the Russians only had 136 kilograms in their tanks. Another reason was Soyuz had no windows to see out, only a periscope. Using techniques originally developed in the Gemini missions, Apollo began the final approach over the Pacific and sidled up to within 30 metres of Soyuz, on its 36th orbit. As they were leaving the Atlantic over Portugal, Capcom Dick Truly in Houston passed up two messages: "Moscow is go for docking and Houston is go for docking. It's up to you, guys. Have fun."

"I'm approaching Soyuz," called Stafford in Russian as Leonov rolled the Soyuz around 60° to help line up the two spacecraft.

"Oh, please don't forget your engine!" replied Leonov in English, meaning Stafford might have rammed him. Everyone in space and at the Mission Control Centers saw the humour and laughed, including a visiting Soviet Ambassador.

"Three metres … one metre … capture … we have succeeded," Stafford announced.

"Well done, Tom. That was a good show. Apollo, Soyuz, are shaking hands now," said Leonov.

The two spacecraft initially locked together at 11:09 am Houston time.

"It was a good show," Leonov sang out happily.

"*Spasebo, Aleksei,*" returned Stafford with satisfaction.

At 11:12:30 the latches were closed to complete the first international docking in space, 998 kilometres west of Portugal, and history was made. Stafford recounts, "We had trained long and hard for several years and when we came up it was a perfect rendezvous – I was speaking to them in Russian as we came in and docked."

Stafford and Slayton were scheduled for the first visit to the Russians. Slayton reached for the docking hatch and smelt what seemed like burned glue. A few moments later it was gone. The Americans had their oxygen masks on standby, but Houston could find nothing wrong, it was probably from an earlier experiment, so they continued equalising the pressure until at last Stafford called out, "Are you ready to open hatch 3?" He then pulled the last door aside at 2:17:36 pm, to reveal Leonov and Kubasov still finishing their preparations.

As the camera cables writhed around the astronauts and cosmonauts in zero-g, Slayton grinned, "Look's like they've got a few snakes in there too."

"Aha," said the Russians. "How are things going? We are very happy to see you." Then Stafford and Leonov shook hands and tried to hug each other.

"Tovarich (Comrade)," replied Stafford as he and Slayton squashed around a table specially set up in to the Soviet spacecraft, and they exchanged gifts of flags and plaques, and some of Leonov's paintings, before standing by to listen to the Soviet Communist Party Secretary Leonid Brezhnev and President Ford's messages. Brezhnev's solemn speech contrasted with President Ford's chatty interview.

Figure 8.5. An artist's impression of the rendezvous. The Apollo Command and Service Module, with docking adapter on the left, approaching the Soviet Soyuz 19 spacecraft.

Ford asked Slayton, "As the world's oldest space rookie, do you have any advice for young people who hope to fly on future space missions?"

"The best advice I can give is decide what you want to do, and then never give up until you've done it," Slayton answered.

"You're a darned good example, Deke, of never giving up," the President pointed out.

President Ford reminded Kubasov of the time they had both gone to a crabfest in Virginia. "We eat good in space, there is some fruit, some juice, and water, a lot of water – no crabs," Kubasov replied with a laugh.

Stafford and Slayton returned to their Apollo spacecraft at around 6:00 pm. The next day Brand visited Kubasov in the Soyuz spacecraft, while Leonov went across to visit Stafford and Slayton in Apollo. Among the activities that followed were television shows, shared experiments, formation flying, joining two halves of medallions, and eating meals together. In the Russian ship they ate strawberries, cheese, sticks of apple and plum, and tubes of Borsch, while in Apollo they feasted on potato soup, strawberries, grilled steak and bread. Leonov had promised Slayton a drink of Vodka if and when they ever met in space. As Vodka was banned from the Russian spacecraft Leonov had filled some tubes labelled "Vodka" with soup, and offered it to Stafford and Slayton, with the lame excuse it was the thought that counted.

The Americans passed across seeds of a hybrid white spruce, while the Russians handed back Scotch pine, Siberian larch, and Nordmann's fir, an ancient Soviet tradition considered a gesture of true friendship. Part of the television shows included narrated views from space of the USSR by both Russians, and of the US by Brand. When asked his opinion of the food Leonov said, "It's not what you eat but with whom you eat that is important."

After 44 hours together the two spacecraft parted over the Atlantic Ocean on Saturday July 21 1975. As Soyuz pulled back Leonov called, "Mission accomplished," and Stafford replied with, "Good show."

Something new for Apollo was the docking module jettison and then its use for a doppler experiment. They first

Figure 8.6. "Tovarich!" Slayton bursts into Soyuz to greet the Russian cosmonauts.

loaded it with rubbish from their Command Module before casting it off on 23 July, and the crew put their suits on for the first time since launch.

Because the Russians insisted on using a frequency commonly used by airport control towers (121.75 MHz), there were moments of confusion while the two spacecraft were docked. At Orroral Valley Tracking Station Ian Fraser commented, "We had trouble finding the spacecraft while they were together, but once they separated we had no trouble finding Apollo straight away."

Once Houston nearly called Apollo up during a sleep period, thinking they were trying to contact them, when really it was an airliner calling the control tower at Atlanta, Georgia. Brand reported, "We've been getting every control tower in Europe and parts of the United States."

Houston Flight Director Littleton adds, "We got a lot of interference over Europe on our television picture and we asked the Soviets to quiet that down. They certainly co-operated on that and turned off some airport communications."

The Americans noticed that the Russian ground control were quite strict with their cosmonauts, more so than the Americans. "We are watching you – we are happy with your work," the Kalinin control centre told the cosmonauts. Both Leonov and Kubasov became tired from lack of sleep and were ordered to take sleeping tablets before going to sleep.

The Russians began their retro fire sequence on July 21. As the Soyuz spacecraft approached re-entry, Kalinin Control Centre ordered, "Report on everything happening, especially during de-orbit. The helicopters and aircraft will attempt to establish communications immediately – so tell them everything that's happening on board."

At 1:00 pm Moscow time many Russian citizens began to gather around television sets to watch their first live Soyuz landing. The automatic re-entry burn was begun about 11,265 kilometres from the landing point. At 9,754 metres the drogue followed by the main parachute ringed with red and white, sprang out, and was spotted by one of the waiting helicopters.

The descent was televised from two helicopters, the signal sent to the United States through the Soviet landlines to Raistings in West Germany, then across the Atlantic through the Intelsat satellite.

At 3,000 metres the heat shield was dropped off then television viewers saw a spectacular sight of the spacecraft disappearing in a cloud of dust and smoke as the landing rockets burst into life 2 metres above the ground to slow the spacecraft to a speed of 3.6 kilometres per hour. The area was soon swarming with over a 100 people and their trucks from the surrounding farms, as the cosmonauts were transported off by one of the nine Mi-8 helicopters hovering around. Leonov said the flight was, "… as smooth as a peeled egg."

The Americans in Apollo continued on, busy with endless scientific experiments until 24 July, when the Service Propulsion System (SPS) rocket was ignited for re-entry over the Indian Ocean, before the Command Module and Service Module separated. The prime recovery ship USS *New Orleans*, stationed 458 kilometres northwest of Honolulu, stood by and waited for the last splashdown – the last water landing. Future spacecraft such as the Shuttle were to land on runways.

Figure 8.7. "So, what do you think of my drawing?" In the Soyuz spacecraft, the Russian Commander, artist, and first man to walk in space, Aleksei Leonov, floated his sketch of Tom Stafford in front of the television camera.

Edgar remembers, "During that final orbit, at the re-entry stage for the CSM the main worry we had was what the predicted velocity was going to be …"

Fraser: "… whether the antenna was going to keep up with the speed of the spacecraft. It was so close it was like you could throw stones at it …"

Edgar continues, "… you could almost go outside and watch it – you could actually see it – and it was daylight. Houston told us, 'we don't think you will be able to keep up with it, we're not scheduling data off this re-entry – but should you get it, we'll take it.' To our surprise we did maintain it…"

Fraser: "… the antenna was flat out …"

Edgar: "… it must have been to design limits and didn't break … right through to the horizon. We heard them talking but it started to break up …

Fraser: "… that's right, the signal started to break up as they came down, but the telemetry stayed okay. We still had the voice on the net (voice line from the USA) after we lost them over the horizon, as they were coming in over the tracking ships and the parachutes were deployed, and suddenly everything went quiet. I think the last thing I remember hearing on the net was coughing before somebody pulled the plug."

Drama stalked the final moments of the last Apollo flight. Just after the spacecraft had plunged into the Earth's atmosphere, shedding kilometre-long flames, the astronauts were suddenly wincing to a loud squeal in the capsule intercom. "The interference was so loud we had to take our masks off and yell at each other," Stafford said. During that time Stafford called Brand to turn on two switches to activate the automatic landing sequence but Brand didn't hear the instruction, and failed to turn the switches on.

When the drogue parachutes didn't appear on time, Brand realised the error and promptly activated them manually, the only time they weren't activated automatically in an Apollo mission. But everyone forgot to turn off the capsule attitude control system, technically called the RCS, so when the spacecraft began swinging around under the parachute, the control system automatically squirted gas out of the thruster jets to try and steady it. Stafford spotted the thrusters were still firing and shut the system down, but the residual nitrogen tetroxide gas was still smoking out of the thruster jets when a ventilation valve opened up at 7,300 metres to let the fresh air in and equalise the air pressures. The

Figure 8.8. "Na Zdrovye!" Stafford (left) and Slayton with the Borsch (soup) containers with vodka labels pasted over they used to toast each other.

lower pressure inside the spacecraft sucked these highly corrosive fumes into the cabin and the astronauts began coughing violently, while their eyes began to burn and sting.

Glynn Lunney, Technical Director for the mission said, "If that gas becomes concentrated enough, it can be fatal." The crew recovered from the surprise and at 2,700 metres manually released the main parachutes.

During the four-minute descent they fought for air to breathe, trying to avoid the acrid brownish yellow gas. After hitting the sea like a ton of bricks the spacecraft fell upside down on the back of a wave, and the astronauts found themselves hanging in their straps as the spacecraft wallowed in the slight sea. Stafford, still coughing with burning throat and eyes, unstrapped himself and fell heavily onto the nose tunnel of the spacecraft before blindly clambering back up for the oxygen masks behind the couches to pass them to the other two.

Brand had passed out, but revived when Stafford clamped his mask on properly. When the spacecraft righted itself the ventilation system cleared the cabin as Stafford yelled, "Get this – hatch open as soon as possible!" but managed to open it himself before the Navy frogman poked his head at the window. Without thinking Slayton gave him a thumbs up gesture; the frogman signalled back to the helicopter the crew were OK, so no special effort was made to get the astronauts out into fresh air quickly.

Stafford comments, "It was touch and go. The oxygen ran out just as we got upright."

When the astronauts climbed out of the hatch they appeared to be normal, except for rubbing their eyes. It wasn't until they were talking to President Ford on the aircraft carrier's deck that the gas incident was mentioned and press conference was stopped then and there while the three astronauts were bundled off to the ship's surgery for treatment. When the crew felt as though they had developed pneumonia, the later functions and parties were cancelled. They were then taken to Honolulu's Tripler Army Medical Centre for intensive care treatment for the effects of the gas, which can lead to the serious lung condition Pulmonary Oedema later on. Although they had trouble breathing for some days after, all three recovered completely.

The 15th and last Apollo spacecraft to take men into space splashed into the Pacific 6.4 kilometres from the USS *New Orleans* and brought to a close the remarkably successful Apollo era of expensive, expendable spacecraft. The reusable shuttle was to take over the task of sending people into space, but only into Earth orbit.

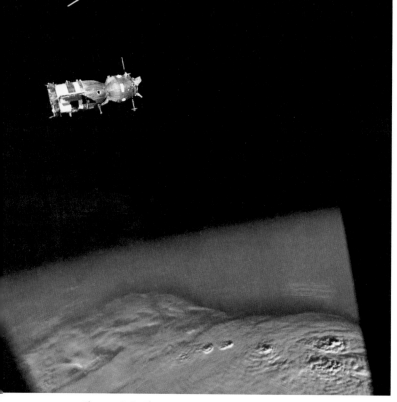

Figure 8.9. The Soviet Soyuz 19 spacecraft from 35 metres as they flew over the northeast corner of South America. The Soyuz spacecraft is made up of three components: the instrument assembly on the left, the descent module with the porthole in the centre, and the orbital module.

The general attitude at Houston was that the flight went off better than expected, there were no major blunders, or accidents, which would have humiliated the space programs in the eyes of the world. Some saw the Apollo–Soyuz as a $250 million political stunt, and felt that the money could have been better spent on another Skylab, but whatever the reaction, it was a congenial way to finish mankind's first steps into space.

However, there was one more event to put the Apollo period to rest – an event which perhaps aptly symbolised the end of the first space era's crashing budgets.

Skylab Bites the Dust in Australia

NASA had expected and planned to keep Skylab operational until the 1980s and the new Shuttle spacecraft could bring it to life again. Unfortunately by February 1978 Skylab's days became numbered when an increase in sunspot activity warmed up the Earth's atmosphere and caused it to expand and reach out for the orbiting laboratory.

Bill Peters was put in charge of a programme to try and prolong the life of Skylab. "We thought that if we could stop it tumbling and lower the drag it would stay in orbit longer. We went out to the Bermuda tracking station in March 1978 as it was the only place that still had the old Skylab ground equipment. There was a guy at Patrick Air Force Base looking at it visually and he could tell when the solar panels were facing the Sun. He would call out (on a phone line), 'It's dark – okay now the wings are facing the Sun,' and I sent a command to turn it on and the RF down-link came on immediately. But as soon as the solar panels looked away from the Sun it shut down. We only had the solar cells working – no batteries.

"Having found we could actually get the thing to power up we came back to the Control Center at Houston and in less than thirty days we designed and built new computers and software and started operations. First of all we had to charge the batteries. The batteries were turned off as they had special circuits that stopped you using the batteries if the voltage dropped below a certain level. Each time we tried to turn the batteries on we could get a few milli-amps of current into the batteries before the circuitry turned it off again. So we just kept sending 'ON' commands – we sent thousands and thousands of 'ON' commands through the tracking stations at Bermuda and Madrid, which came on line later. We spent April and May charging batteries until eventually we got them all up to the right voltage. I remember sitting there looking at the strip pen recorder which gave a bleep when the Sun sensor came on. At that moment I sent the command, I think it was through Madrid, to hold attitude. We didn't stop it tumbling until about June, when we could enable the gyros to hold the lab in that attitude."

The Shuttle was still struggling to get off the ground, so as Skylab began to approach the first wispy tops of the Earth's atmosphere the increasing resistance began to drag it down until there was a new predicted re-entry in June or July 1979 when it was expected to end its life disintegrating into a molten fireball – but where would it come down? Bill Peters says, "They had a series of six orbits for it to come down in at the right altitude and as fate would have it we didn't have to do anything to get within those six orbits."

At NASA headquarters the staff went onto 24 hour duty during the three day Skylab "death watch", beginning on Sunday 8 July. The Johnson Space Center, the Marshall Space Flight Center, and the Kennedy Space Center stood by. Telephones began to ring incessantly with serious, funny, and frightened people wanting to know what was happening and complaining their holidays were ruined. The news media moved in, and by Tuesday there were representatives from the White House, the Federal Preparedness Agency, and the Departments of State, Justice, and Defense set up in the NASA offices. News bulletins were issued every six hours, then every hour as Skylab plunged back to Earth in its final death throes over the Indian Ocean on Thursday 12 July 1979.

They began 111 kilometres over Ascension Island in the Atlantic when the radar station there spotted the big solar panels begin to tear off as the lifeless hulk spun and twisted out of control. "It's now out of range of all our tracking stations," said NASA. "The crash line is from Esperance in Western Australia to Cape York in Queensland. The chances of anybody coming to harm are minimal, but people are advised to stay indoors."

During Skylab's last week in space, the Australian Federal Government set up a special Skylab Communications Centre in the Deakin Telephone Exchange in Canberra. Manned by about 12 officials from 5 departments, it monitored every move Skylab made over a hot line from Washington. Police and emergency services around Australia were put on alert. People all around the Earth under its flight path nervously wondered.

In the United States all aircraft in the north-eastern and north-western areas were grounded as Skylab passed overhead for the last time. 400 members of the world's media had gathered at NASA Headquarters in Washington where a statement was issued that Skylab had come down safely in the Indian Ocean, calculated from the last radar tracks.

Some celebrations had already begun in America for the safe ending of Skylab.

Then, quite unexpectedly, there were disjointed reports from around the desert 800 kilometres behind Perth. "There have been reports of sightings of fragments over Australia – from Kalgoorlie, Esperance, Albany and Perth," NASA officials announced. In the middle of winding up the story on the end of Skylab the journalists at NASA headquarters in Washington were electrified into action: "Where's Albany?"; "How do you spell Kalgoorlie?"; "Where's this Perth?"; and suddenly the sleepy little outback towns of Kalgoorlie, Albany, Rawlinna, and Balladonia were thrust into the world's major newspaper and media headlines.

Captain Bill Anderson was flying his Fokker Friendship 200 kilometres east of Perth on his final approach to Perth airport when his first Officer Jim Graham saw a blue light through his left window. Anderson recalls, "We first saw it at 12:35 local (Perth) time – we would have watched it for about 45 seconds. I had the impression it was a bubble shape. As it descended it changed from a bright blue to an almost orange-red and you could see the breakup start to occur. It finished up as a very bright orange ball in the front, and the remainder behind giving off sparks. It was a very long tail, perhaps several hundred miles long."

Bradley Smith, an employee at Perth's Bickley Observatory, described his sighting. "We first saw it as a light behind the clouds. It was travelling from south to east about 9° above the horizon. If you can imagine a train on fire with bits of burning fire all the way down the carriages – that's what it was like."

John Seiler, managing the remote sheep and cattle station of Noondoonia 850 kilometres east of Perth saw the final moments of Skylab with his wife Elizabeth. "I was watching for it – and saw it coming straight for us. It was an incredible sight – hundreds of shining lights dropping all around the homestead. They were white as they headed for us, but as they began dropping the pieces turned a dull red.

"The horses on the property ran mad. They galloped all over the place, and the dogs were barking. We couldn't calm them down. Then we could hear the noise of wind in the air as bigger pieces passed over us – all the time there was a tremendous sonic boom – it must have lasted about a minute. Just after the last pieces dropped out of sight, the whole house shook three times. It must have been the biggest pieces crashing down. Afterwards there was a burning smell like burnt Earth."

NASA officially revised its re-entry bulletin to: "Skylab re-entered the atmosphere at an altitude of 10 kilometres at 2:37 a.m. (Eastern Australian time) at 31.8° S and 124.4° E – just above the tiny Nullarbor Plain town of Balladonia." Burning pieces of Skylab were scattered over an area 64 kilometres wide by 3,860 kilometres along the flight path.

President Jimmy Carter sent a message to the Prime Minister of Australia, Mr Malcolm Fraser:

Dear Malcolm,

I was concerned to learn that fragments of Skylab may have landed in Australia. I am relieved to hear your Government's preliminary assessment that no injuries have resulted. Nevertheless, I have instructed the Department of State to be in touch with your Government immediately, and to offer any assistance you may need.

Sincerely,
Jimmy Carter

Prime Minister Fraser's reply:

Dear Jimmy,

Thank you very much for your message. It appears we can all breathe a sigh of relief. While receiving Skylab is an honour we would have happily forgone, it is the end of a magnificent technological achievement by the United States, and the events of the past few days should not obscure this.

If we find the pieces, I shall happily trade them for additions to the beef quota.

Warm personal regards,
Malcolm Fraser

Skylab, made up of Apollo Moon mission leftovers, took 6 years, 1 month, and 27 days to travel 878 million miles after 43,981 orbits of the Earth. Dr George Mueller's dream ended up as a few charred and twisted souvenirs of metal and fibreglass recovered by locals and tourists of the outback of West Australia, briefly setting off a mini boom in fragment hunting as NASA, and other organisations reportedly offered rewards of thousands of dollars for the biggest chunk of Skylab. Some of the wreckage found went on display in museums and institutions around the world. One piece of aluminium, thought to be a door weighing 82 kilograms, was found near Balladonia.

The demise of Skylab brought to an end the first steps to send Americans into space. Skylab was the last operational item left from the Apollo days. It chose to end its life between the original Australian Project Mercury stations of Muchea and Red Lake – full circle from where the whole adventure had begun from Australia's point of view, just under twenty epoch making years before.

The First Human Burial on the Moon

A legend among geologists, Eugene Merle Shoemaker was born in 1928 and died in a motor car accident in 1997 in the Australian outback near Alice Springs while searching for meteor craters. He was synonymous with the early Apollo geological activities, in fact, going right back to 1948 when he first heard about the V-2 experiments at White Sands Proving Ground. He guessed that America would be exploring space and immediately decided that he wanted to be the first geologist cracking rocks on the Moon. He kept these dreams to himself as in those days few believed humans would ever go to the Moon, and such ideas were usually regarded with derision. Later when asked where he would like to be in twenty years time he would wave a hand at the Moon and say, "Up there." To fly to the Moon as an astronaut with his own hammer was always his main ambition until he was diagnosed to have Addison's Disease in 1963.

Shoemaker always seemed to be pursuing new ideas, prodding his calculator, and worrying over theories not working out. He had a very persuasive character, probably because he was always passionate about his subject and knew what he was talking about – when he talked people listened and found the experience mind expanding. He had the ability to entertain his listeners with a very boring subject. These traits in his character were to help the geologists cause with institutions such as NASA.

In 1956 he became a founding father of the United States Geological Survey's original programme of lunar investigations and his vision led directly to the lunar field work carried out by other geologists and a group of geologically trained astronauts.

In 1959 Shoemaker began a year's intensive study of Meteor Crater on Coon Mountain in Arizona to determine exactly how impact craters were created to better understand the lunar craters and how they were formed. From his studies of Meteor Crater he was the first person to realise the awesome destructive power and worldwide implications of space debris hitting the Earth. He lived to see one of these collisions between the planet Jupiter and that of his comet Shoemaker–Levy in 1994. Geologist/author Don Wilhelms noted, "Few individual scientists have contributed so much of fundamental importance as Eugene Shoemaker did in 1959 and 1960."[14]

From Arizona's Meteor Crater Shoemaker swung his energy over to studying the large lunar crater Copernicus. Using a detailed photograph taken by Francis Pease in 1919, he concentrated on the ballistics of crater ejecta and the smaller craters that surround all young and many old large craters – the satellitic craters. This study led to his creating the second modern lunar geological map. This map marked the birth of a systematic lunar-geologic mapping program that was carried out by the US Geological Survey for the next two decades and continues today in the more general form of planetary mapping.

Next he published a paper which laid the foundation for all subsequent studies of the crusts of the Moon and planets based on historical concepts.

In early 1963 Shoemaker, representing the US Geological Survey, worked out an agreement with Maxime Faget of NASA's Mission Control Center to train the astronauts and provide geological support such as instrument development and mission simulation.

Also in January 1963 Shoemaker became the principle scientific investigator of the television experiments of the Surveyor soft landing project, which became an official US Geological Survey project in 1964.[17]

Shoemaker later turned against NASA and the Apollo Program. He believed that sending people into space was for making discoveries, and felt the potential of the Moon Landing program was wasted, and blasted NASA for "a miserable job" integrating science and scientific goals into the Apollo missions.

"Not going to the Moon and banging it with my own hammer is the biggest disappointment in life," he once said, so when Carolyn Porco, a colleague and planetary scientist at the University of Arizona, read of his death in the papers she immediately thought, "… how badly he had wanted to get to the Moon all flashed through my mind with the speed of a cosmic impact. Let's send Gene to the Moon, I thought. This is his last chance."

After getting approval from Carolyn Shoemaker, his wife recovering in hospital in outback Australia, it took Porco only 30 hours to set the wheels in motion to add Shoemaker's ashes to the Lunar Prospector spacecraft, at the time undergoing final systems testing.

The aptly named Lunar Prospector left Earth carrying Shoemaker's ashes on 6 January 1998 and circled the Moon for 18 months before hurling itself into the lunar surface near the South Pole at 6,115 kilometres per hour, kicking up clouds of dust and dirt. Thus, on 31 July 1999 Eugene Shoemaker was buried where he had always wanted to be. Shoemaker's friends and colleagues said, "This is very important to us. We will always know when we look at the Moon that Gene is there."

We, and future generations can also look up to the Moon riding across our skies to always remind us of the Twentieth Century days of the Apollo era, when humans first successfully left the cradle of the Earth to walk and explore another celestial body.

To put this into perspective, consider that it takes 225 million years for our solar system to travel around the spiral Milky Way Galaxy just once. If the human race in any form has been in existence for even 3 million years, we have only just begun a fraction of the trip around our galaxy for the first time – and in that brief moment of time we have already left our home planet. So, looked at from a galactic perspective it would seem a very impressive achievement indeed.

Should anyone question the wisdom or reasons for the Apollo Projects, perhaps we should go back to the beginning and quote Benjamin Franklin. When he was in Paris watching the early balloon flights he was asked, "What good are they?"

His reply? "What good is a new-born baby?"

Figure 8.10. The next step. ▶

Appendix 1

Manned Spaceflight Costs

Including the rockets, spacecraft, tracking, operations and facilities, in United States Dollars:

Mercury Program	$392,600,000
Gemini Program	$1,283,400,000
Apollo Program	$25,000,000,000
Skylab Program	$2,600,000,000
Apollo Soyuz Test Project, US only	$250,000,000

1981 The End of ALSEP

The ALSEP scientific packages left behind on the Moon by the astronauts were turned off permanently on 1 October 1981, and Houston advise they are unable to be turned on again. However there are some passive elements such as laser mirrors which can still be used at any time into the forseeable future.

1996 Lunar Sample Locations

The 12 moonwalkers spent a total of 300 hours on the moon and brought back 2,196 samples of lunar rocks and soil weighing a total of 381.7 kilograms. Eighty percent of the materials are stored in the Lunar Sample Building at Houston's Johnson Space Center. Five percent are on loan to scientific institutions, schools and museums. Three percent have been destroyed by testing. The remainder is stored at Brooks Air Force Base in San Antonio, USA.

Three samples of Lunar material are on display in Australia: one at the Power House Museum in Sydney and two at the Tidbinbilla Tracking Station near Canberra, with a flag that accompanied the Apollo 11 astronauts to the Sea of Tranquillity.

The three automated Soviet missions in 1970, 1972 and 1976, brought back 11.3 oz (320 grams) of lunar material.

Figure A.1. A map of the Moon showing the main landing points of the American and Russian missions. If viewing from the Southern Hemisphere, the map should be turned upside down.

Appendix 2

An Interview with Dr Stuart Ross Taylor

Dr Stuart Ross Taylor, MA, DSc (Oxford), MSc (NZ), PhD (Indiana), FAA, is a Professorial Fellow at the Research School of Earth Sciences at the Australian National University, Canberra.

He was a member of the NASA Lunar Sample Preliminary Examination Team for Apollo 11 and 12. He was put in charge of the initial chemical analyses and carried out the first chemical analysis of a lunar sample. He is a Principal Investigator for the Lunar Sample Program, a Fellow of the Meteoritical Society, and of the Australian Academy of Science, and a Foreign Associate of the National Academy of Sciences of the United States of America. He has published 210 scientific papers and six books.

The author interviewed Dr. Taylor at his home in Canberra in May 1994, for an appendix on the results of the Apollo lunar sample analyses for this book.

HL: "Dr Taylor, can you tell us how you came to be involved in the Apollo 11 Lunar Sample Analysis Team?"

Taylor: "I happened to be in the United States early in 1969 at a conference on tektites – we were having one of those ding-dong battles about tektites being lunar versus terrestrial. In fact we took a vote – this was in March '69 – and my New Zealand colleague Brian Mason, the person who I had done my PhD with, who is a meteorite expert and the Chairman, called for a vote, and it was 50:50 – this was just four months before the mission.

"Then I went down to Houston, where another Australian, Robin Brett, was the Chief Geochemist there. It's interesting the number of people from the British areas that were involved in the Apollo missions – the place was full of them.

"Robin said, 'We have this spectrographic laboratory running here, and we're having troubles with it, I want you to come in and run it.' I took rather a deep breath at this, since I had left my wife and family in Canberra building a house, this is the house we're sitting in now, so I called her up, and I stayed six months in Houston running this laboratory.

"It was one of the most difficult periods of my life because it took every bit of experience and skill I had to get it running and get the answers out. For example when we got the first lunar sample in our hands, we had a press conference four hours later at which we had to produce results."

HL: "You were on your mettle there!"

Taylor: "You're right – the searchlight was right on you!"

HL: "And you had to be right!"

Taylor: "And you had to be right, yes!"

HL: "Can you describe what it was like in the Lunar Receiving Laboratory?"

Figure A.2. Dr Ross Taylor at the Spark Source Mass Spectrometer analysing lunar samples at the Australian National University in Canberra. "This machine was like an old friend. I spent many hours, first back in the 1960s, adapting the machine to analyse rock samples – when the lunar samples appeared, we had just solved several difficult analytical problems."

Taylor: "Well, it was quite strange, because it was isolated behind quarantine barriers, so when you went in, you had to get out of all your clothes, and change into what looked like a surgical suit. The samples themselves were protected behind biological cabinets where one used rubber gloves. The cabinets were held with a positive pressure of nitrogen. We did all our analyses through these gloves which made it quite difficult technically to handle this material, specially as these lunar samples all had a static charge on them. We felt the quarantine was unnecessary because the chances of the Moon having anything harmful biologically was very remote, but, of course, you can't be totally certain.

"The real problem was that the quarantine had been breached in the middle of the Pacific Ocean when they opened the hatch and got the astronauts out – they were covered in lunar dust, so already at that point the Earth's surface had been exposed to the lunar material so, you know, if the Pacific Ocean had turned bright green … so we felt it was more of a public relations exercise than anything."

HL: "The American Nobel Prize winning chemist Harold Urey was one of the first driving forces to initiate the Apollo Moon missions. Did these missions bring home the results that Urey had hoped for?"

Taylor: "Well not exactly. Harold Urey had expected that the Moon was a primitive object left over from the very beginning of the solar system, and he figured out that by going to it we would be able to look at this very primitive object and work out what had happened at the beginning of the solar system. It turned out of course that the Moon was a highly complicated object with a geological history of its own, and that, somewhat ironically, it did provide us with keys to what went on in the early history of the solar system mainly by highlighting the importance of large collisions of meteorites and asteroids with the Moon, but certainly not in way that Harold had envisaged."

HL: "Are all scientists in agreement with the results?"

Taylor: "Yes. It's fair to say that there is a consensus view now on the origin and evolution of the history of the Moon basically from the results of the Apollo missions and the reason for this is that all the other alternative theories have pretty much been demolished."

HL: "Is it true that we know more about lunar rocks than some terrestrial counterparts?"

Taylor: "Yes. Again there is a certain irony in this. I think we understand both the origin and the evolution of the Moon more clearly than we do that of the Earth, partly because the Earth is a much bigger object, and we are still uncertain about quite a lot of the details of the interior of the Earth. The Moon is very much smaller – it's only one-eightieth of the mass of the Earth, and it has had a very much simpler history. For these reasons I think we have a very much clearer view of the history and origin of the Moon."

HL: "Can you comment on the statement, 'Never in the field of human endeavour was so much discovered by so few from so little.'"

Taylor: "Well, of course, this goes back to Churchill's famous comment on the Battle of Britain. It's certainly interesting in that it has taken two or three hundred years of geological exploration of the Earth to sort out what has happened. It's only in the last twenty or twenty five years that we understood, for example, that plate tectonics was operating in the Earth, and this was the result of an immense effort by generations of geologists and geophysicists. On the Moon we had it sorted out in about a decade."

HL: "Have there been any common events found that can be connected between the Moon and Earth?

Taylor: "Well, yes in the sense there are meteorite impacts on both bodies, and on the Earth the evidence of these events is pretty much wiped out by erosion and weathering and so on of rocks. So the evidence of these large impacts was very hard to find in the Earth and was doubted for many years by geologists, whereas the Moon is covered with craters, there being no atmosphere and no weathering – if a crater forms on the Moon by an impact, it stays there until it's wiped out by another impact. On the Earth these things are eroded away, so for many years geologists doubted that there were such things as large meteorite impacts on the Earth. Now the evidence has become quite clear, we have found the remnants of old craters. So perhaps this is one of the few things that the Earth and the Moon have in common geologically.

HL: "The Moon is riddled with rilles – are these rilles tunnels that have collapsed?"

Taylor: "Well, they caused a lot of questions. Here are these channels looking a little bit like river channels. They have obviously been cut by some kind of fluid, and so people thought this was evidence of water on the Moon. However, it turned out that these rilles were cut by lavas and they're really quite big – Hadley Rille, close to Apollo 15, is about a kilometre across, and these are much larger than lava channels in the Earth. The reason for this that the lavas are very much more fluid than on the Earth because they turn out to have a much higher iron content, so they have the viscosity of about engine oil. They flow very easily compared to the more viscous lavas of the Earth which are almost like toffee. So these lunar lavas have the properties almost close to that of water, so they are able to cut these channels. They are probably open channels. It would be very difficult to roof over something a kilometre or so in width, so the feeling is they were always open or perhaps had a skin of lava over the top. We still don't have a complete answer to this question."

HL: "Two questions about the lunar soil. Would it have a smell?"

Taylor: "Not to my knowledge. I certainly have handled plenty of it but I suspect there is nothing in it to give rise to a smell. It is totally dry and there's no organic material. I think smell mainly comes from the volatile organic materials or molecules which our noses are very sensitive at picking up."

HL: "You see, the astronauts mentioned a smell when they climbed back into the LM?"

Taylor: "Yes. I've got no experience in that really."

HL: "So you think they might have smelled something from the descent engine gases or something inside the LM?"

Taylor: "It's possible."

HL: "There wouldn't be any interaction with the oxygen from the interior of the LM?"

Taylor: "No, I don't know. There was none fortunately."

HL: "If lunar soil came near water, being so dry – or not necessarily water – but something that had moisture on it, would it tend to leap up to grab the moisture?"

Taylor: "It was so dry it had this static charge in it and so you could put the lunar soil into a small vial and you could tip the vial upside down and it wouldn't fall out because it would stick there. This was because of the static charge on this extremely dry material."

HL: "Is there any sign of moisture on the Moon?"

Taylor: "We did not find a drop. The Moon is dry as a bone. There is a feeling that there might be, in some of the craters at the south pole, which are in permanent shadow. There might be ice there, perhaps as a result of cometary impact, but the Moon itself has no water on the surface, no water in the rocks, no bound up hydroxyl in the minerals. All the minerals are anhydrous, unlike on the Earth, where we have many minerals which have water or hydroxins in their structure."

HL: "What was the reaction when lunar soil came in contact with terrestrial water?"

Taylor: "Well, none that I am aware of. In the lunar soil there are quite a lot of small glassy fragments from the meteorite impacts and so on. These tend to contain elements – small amounts of potassium – in these glasses, and these are leached out by the water so that it made the lunar soil quite fertile for things to grow in."

HL: "What were the results of the trials growing terrestrial plants in lunar soil?"

Taylor: "They were really quite spectacular – at first! This caused a lot of interest in the Lunar Receiving Laboratory. The potassium was leached out rather easily from the glass. The plants grew quite well in it. Then when we checked what these botanists were doing, their control soils were quartz sand, and, of course, quartz sand is a pretty barren medium for things to grow in, so the plants growing in the lunar soil grew much better than the plants growing in the quartz sand. So it looked as though soil that was enriched would be a great thing to grow plants in. But the experiment was not very well controlled."

HL: "Does the Moon have a solid, or a molten core?"

Taylor: "It's unfortunate that we did not have a decent meteorite impact on the far side of the Moon in order to give us a seismic signal with which we could map the centre of the Moon properly. It turns out that the centre of Moon is possibly liquid, but we don't really know that. There's almost certainly a small iron core of perhaps about two or three hundred kilometres, about two percent of the mass of the Moon, a very small core in the centre, but that remains one of the uncertain areas which we would hope to find out about in future missions."

HL: "The Apollo missions left a number of ALSEPS (equipment for science experiments) on the Moon. One of the experiments used a seismometer. Can you tell us some of the results of the seismic activity from the Apollo ALSEPS?"

Taylor: "Well they detected a series of small moonquakes, very low in intensity, and most of these repeated on a monthly cycle as the Moon rotated around the Earth, and it seemed to be these were just basically creaking and groaning of the Moon, if you like. Tidal stresses in the Moon due to its rotation, not due to internal processes, as we have on the Earth with earthquakes. They were just sort of tidally induced stresses on the Moon. The ALSEPS did, of course, pick up the impact of a large meteorite, about a one tonne object, on the far side in 1972, which gave us some information about the interior of Moon."

HL: "At the tracking station we could see the surface cracking and moving at sunrise as the temperature rose and the sunlight crept across the dust of the lunar surface. How deep did this activity go?"

Taylor: "I think probably only a few centimetres at most. The surface is covered with a layer of rubble anywhere from two or three, perhaps up to ten metres deep, of ground-up, smashed-up rock, and boulders and fine soil, all continuously being reworked by meteorite impact. This is the so-called regolith, and this acts as an insulating blanket over the surface. Any temperature effects would be only in the upper few centimetres."

HL: "Can you explain why the Moon rings like a bell. We used to hear about this quite a bit during the days of Apollo when we used to crash the LM onto the surface?"

Taylor: "Yes, well the seismometers picked this up – this was one of the extraordinary things. When you had the impact of these objects on the Moon, unlike a terrestrial earthquake which dies away quickly, the shock waves would continue to reverberate around the Moon for period of an hour or more, and this is attributed to the extremely dry nature of the lunar rocks. There is no moisture on the Moon, and nothing to damp out these vibrations. The Moon's surface is covered with rubble and this just transmits these waves without them being damped out in any way as they are on the Earth. So they just ran around the Moon, basically. It's a consequence of the Moon being extremely dry."

HL: "We had Apollo 11 through to Apollo 17. Did we change our experimental attitudes much between 11 and 17 – in other words did we learn much from Apollo 11 to incorporate into the future experiments?"

Taylor: "Well yes, because a lot of what came back from Apollo 11 was unexpected, or at least there were many different schools of thought about what we would find on the Moon, in particular the age of the lunar surface and the composition of the rocks. There were a large number of surprises from the initial Apollo 11 mission about the composition of the rock, and the age of the rocks. This wiped out a lot of the pre-Apollo speculation, and led us to focus in on quite new directions on what had happened to the Moon. For instance, it had not been thought that the Moon might have been molten early in its history, and yet this was one of the things that came out of the early work. So we had to re-orient our thinking really quite dramatically. Nothing like a few hard facts to come back to really dispose of a lot of wishful thinking."

HL: "Which Apollo mission had the most interesting samples?"

Taylor: "I think it was Apollo 15."

HL: "That was the one at Hadley Rille and the Apennine Mountains."

Taylor: "Yes, that's right – because it was a good site to go to. It had both the maria and the rim basin, the rille and the highlands. Apollo 17 was a bit of a repeat of it in some ways.

HL: "Were any useful minerals found?"

Taylor: "Well it depends on your definition of useful, of course, but there are refractory minerals. There is quite a lot of feldspar, a lot of titanium oxides, elements like this, but none of these would be particularly useful economically on the Earth – they are very common elements. It is ironic that the elements that we think of as precious, pink diamonds for example, or rare like the platinum group elements – lead, silver, or gold, – these sort of elements are all very low in abundance in the Moon which is partly a consequence of the way the Moon originated and its subsequent history. There is a certain irony is this. For example, some people thought that the Moon's surface might be covered with diamonds, because there would be carbon there, and with the meteorite impacts – the high pressure would produce diamonds. Unfortunately there is no carbon, so that almost everything we think of as an ore mineral is in fact missing on the Moon. There is one thing present, helium 3, which comes from the solar wind, which could be a mineable resource because helium 3 would be a very useful fuel for fusion reactors on the Earth."

HL: "Can you explain why lunar rocks are much older than terrestrial rocks?"

Taylor: "Because the geological history of the Moon started right back when the Moon was formed. The Moon was melted, it crystallised and formed a crust. Lavas came up from the interior – all this happening about 3 or 4 billion years ago – and then the Moon shut down, so that most of the rocks on the Moon are significantly older than the rocks on the Earth. The Earth goes through many cycles of erosion and material is removed and recycled back into the mantle. So there's a cyclical process going on in the Earth which is not going on in the Moon. The net result is that most of the rocks in the Earth tend to be very much younger than those we see on the Moon.

HL: "Dr. Taylor, from what you have told us, there is not very much on the Moon to support life. Can you comment on that, please?

Taylor: "Yes. The basic problem, of course, is total lack of an atmosphere and lack of any oxygen. There is also a lack of carbon in the soil. There is nothing to build life from as we know it. The surface is exposed to extremes of heat, cold, and hard vacuum – a much better vacuum than we can get in a terrestrial laboratory – also the absence of water. The other problem is it is directly exposed to solar radiation, so all the ultra violet radiation from the Sun falls directly on the surface. It's almost an ideal sterile environment."

HL: "If there is so much carbon on the Earth, why is there none on the Moon?"

Taylor: "One would have expected any body to have carbon in it. I think the point is that during the event it was cooked up to such a high temperature the carbon just didn't condense back into it. The carbon was lost, along with water and other things. It's the same argument why there is no water on the Moon."

HL: "It was thought that the spacecraft might be swallowed up by a talcum like dust. Do you think that there are any pockets of dust bowls such as this?"

Taylor: "No, but on the surface there is this layer of rubble which is boulder size down to quite micron size particles which have been smashed up by the meteorite impacts. There is quite a lot of fine dust mixed in with this, but the idea went back to Tommy Gold, the astronomer, who had suggested that the lunar maria were in fact gigantic dust bowls, filled up with dust from the craters in the lunar highlands, and these were perhaps two or three kilometres deep. Any spacecraft landing on these maria would sink straight into this dust. It cost the American taxpayers a lot of money to send up Surveyor missions to establish the bearing strength of the surface. The Surveyors had a little shovel to scrape away the surface, and cameras to look at the landing pad. They found that the bearing strength of the surface was about the same as that of a sandy beach, so the spacecraft would only sink in a few inches."

HL: "What are the highest and lowest temperatures recorded on the Moon's surface?"

Taylor: "Well, at the Apollo sites the maximum temperatures were up to about 110°C which is above the boiling point of water, and the coldest temperatures are down to –170°C. This, in the absence of an atmosphere and convection, if one side of an object is illuminated by the Sun it is hot, and the other side in the shade would be freezing cold. This led to the reason for wrapping the legs of the LM with gold to get good conduction of heat from one side to the other to avoid thermal stresses on the landing legs, which may have caused them to break on impact."

HL: "Are we any closer to finding out where the Moon came from after the Apollo missions?"

Taylor: "Yes. One of the reasons that Harold Urey had said we were going to find out the origin of the Moon and origin of the solar system was by going there. By about 1984, about 12 years after the landings, very little progress had been made on the question of the origin of the Moon, and it became a little embarrassing when people would ask us, 'Well, you have had 12 years to work on these samples. Surely you can answer the question where the Moon came from?'

"People have really been very busy working on the detail, the chemistry, the ages, and all the other scientific investigations, and to some extent the origin of the Moon had been put in the too hard box, because all the pre-Apollo theories of capture, fission, or double planet, collapsed with the lunar data. None of the Apollo data fitted any of these preconceived theories and it wasn't until about 1984 that a conference was held in Hawaii. It was thought at this time we had better have a conference to sort out this problem. You know, when you have an unsolved problem you have a conference. Then they thought, well, because of the old idea that the Moon came out of the Pacific Ocean, an idea going back to George Darwin, it might be some special insight might be gained by holding the conference in Hawaii in the centre of the Pacific Ocean.

"So we held it there, and in fact we did get some special insight – this was the idea that the Moon had originated as a result of a glancing impact of a body the size of Mars hitting the Earth in the very early days, just as the Earth and the other planets were being formed. A lot of large collisions were appearing and we happened to have a chance collision with a Mars sized object that then spun out of the rocky mantle to form the Moon. This explained a whole host of observations – first of all why the Earth/Moon system is spinning rapidly around. Something had given it a very large kick, and the other reason, why the Moon is rocky and Earth is much denser, as well as the strange orbit of the Moon … and the chemistry as well."

HL: "While on the subject of collisions, I presume you've found nothing to support Velikovsky's theory of a passing planet upsetting the Earth's equilibrium?"

Taylor: "No. There have been some computers designed to look at the long term stability of the solar system and these are called orrerys after the mechanical models of the orrerys which are the mechanical models of the solar system built a couple of hundred years ago. These are computer versions of the orrery looking at the long term stability, and it seems from this that there's no sign that the present arrangement of the planets has changed over the past 4 or 5 billion years – since the solar system settled down into its present arrangement.

HL: "Has there ever been any volcanism on the Moon?"

Taylor: "Yes. There's quite a lot of volcanism, but not exactly in the sense we understand it on the Earth, in that you don't get Fujiyama type volcanoes on the Moon. The dark patches that form the features of the man in the Moon, are actually lava flows, but as I said earlier, they are extremely fluid so they flood out over the surface of the Moon filling in all the hollows. They melt in the interior and come up as lavas do on Earth, then they just flow widely over the surface because they are so fluid, and so they cover very large areas, but they are mostly flat lava plains and we do have examples of this on the Earth. The conventional conception of a central volcano is like Hawaii, Fujiyama, or Mount St Helens. These volcanoes don't exist on the Moon. The large craters we see are all due to meteorite impact.

HL: "When did the last volcanic activity occur on the Moon?"

Taylor: "About three billion years ago, perhaps a little later – it could be two and a half billion years."

HL: "What is the age of the oldest sample that Apollo brought back?"

Taylor: "There has been a debate about that. Measuring the age is quite a difficult procedure to do, which is why you really have to do it in a terrestrial laboratory. I think the oldest genuine age is 4,440 million years with an error of about 20 million years on it. This is the age of the white lunar highland crust. This was only about 100 million years after the formation of the solar system. So the Earth and the Moon had to be formed in a fairly brief period of time. 100 million years after the formation of the solar system, it was looking pretty much as it does today."

HL: "How does this compare with the age of the Earth's surface?"

Taylor: "Well the oldest rocks we have on the Earth's surface are just under 4 billion years old, and these are very tiny remnants of rocks in the northwest territories of Canada. So most of the old rocks on the Earth's surface have been eroded away. There were obviously rocks present at that time, but most of these have disappeared through erosion and recycling. So the Moon gives a us a unique sort of window into the early solar system."

HL: "So when you say recycling you mean going back onto the Earth's mantle?"

Taylor: "Going back in the mantle, or just been destroyed by erosion and so on."

HL: "The displacement of the crust of the Moon away from the Earth. Can you describe what caused this displacement and how the lunar mountains came to be formed?"

Taylor: "This is one of those features which has been known for a very long time, about a hundred years or more, that the centre of mass of the Moon is displaced towards the Earth by a couple of kilometres, relevant to the geometrical centre of the Moon. If you work out where the geometrical centre is, the centre or the mass of the Moon is displaced a little

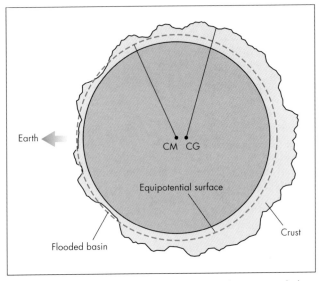

Figure A.3. Cross-section through the Moon in the equatorial plane showing the displacement of the Centre of Mass (CM) towards the Earth relative to the Geometrical Centre (CG).

bit towards the Earth, and the reason for this seems to be the crust of the Moon, which is about 60 kilometres thick on the near side – the white area you see on the Moon, the so-called highland crust. It's basically feldspar, which is why it is white, and on the far side it's about a 100 kilometres, or even a bit thicker. It is rather low density material, so basically the Moon has an excess of low density material on the far side which means the centre of mass is actually displaced a bit towards the Earth, and it's locked in to rotating around the Earth with one side permanently facing the Earth. As the Moon rotates around the Earth, its centre of mass is pulled a little bit towards the Earth.

"One of the striking features of the mountains, which puzzled people from the earliest telescope observations, was the circular arcs of mountains, quite high mountains, up to 15,000–20,000 feet. The fact that these mountains tended to lie along the arcs of circles puzzled people. Then around the impact craters there are quite substantial peaks as well, and sometimes there were central peaks inside the impact basins, which are due to rebound after the meteorite hit.

"But the big circular mountains, the Carpathians and so on, are in fact the result of asteroidal impact, perhaps 50 or 100 kilometres in diameter, hitting the Moon. It's a bit like dropping a stone in a pond. The rocks behave fluidly, and as the meteorite hits, you get these ripples, a series of waves, going out, and these effectively freeze as the mountain reaches 2–3 kilometres high, so that these great structures – one of these big circular structures is the size of France – have several circular rings of mountains around them. It was probably formed in five minutes."

HL: "As the Moon always has one side facing Earth, would one side of the Moon receive more impact craters than the other side? Would the Earth protect the side facing in?"

Taylor: "The locking of the Moon's orbit to the Earth was fairly recent, well, since the time when the very large impact flux was going on. Right back at the beginning the Moon was still rotating relative to the Earth."

HL: "How did the samples brought back from the Moon compare with the Surveyor results – particularly Apollo 12 and Surveyor 3?"

Taylor: "Well, Apollo 12 landed within a few hundred metres of Surveyor 3, which was really a great technical feat."

HL: "One hundred and eighty-three metres."

Taylor: "Yes, 183 metres, okay. It was certainly a remarkable technical feat since Apollo 11 had been off course by four miles or so. Surveyor 3 had chemical analysis equipment on board which signalled back that it was sitting on something looking like a basaltic lava but was rather high in titanium, and people didn't tend to believe this result, but when we got the first analyses from the lunar samples it showed the fact that they were high in titanium and this vindicated the Surveyor 3 analysis over which there had been some scepticism."

HL: "So as a scientist you strongly support obtaining samples as against sending a remote sensing device to analyse soils on other bodies?"

Taylor: "Well, yes. In spite of what I just said about the relative accuracy of the Surveyor 3 analytical equipment. You can certainly obtain certain sort of rough surface analyses and you can do this for the major elements but if you want to look at either the trace element content, or the isotopic abundances, or the age determination in particular, you really have to get a sample back into terrestrial laboratories. It's quite difficult to do the analysis here – it's not that you can't design the equipment. You can make beautiful mass spectrometers and equipment to work on these surfaces, the real trouble is not so much the equipment, but it's what sample you put in, you've got no control over what you put into this equipment.

"We saw this on Mars with the Viking Landers where there were three experiments designed to detect biological activity, or life, and they returned very enigmatic results. Either you get the result you expected, or else you get something you can't interpret. Obviously what seems to have happened to Mars was that the soil was quite strongly oxidising and this gave the simulated effect of life. The real conclusive experiment on the Viking landing was a mass spectrometer which found no trace of carbon so that there was no real possibility of having a carbon-based life present. The biological experiments were so unexpected that we couldn't really interpret them. That's the problem with remote sensing."

HL: "Do you feel that the samples brought back cover the lunar soil in general, or do you feel that there may be older examples, or materials we haven't thought of there?"

Taylor: "Fortunately the Moon is a lot simpler than the Earth. People often say if you brought back a sample from the Earth – if you landed in six places on the Earth you couldn't possibly work out the geological history of the Earth from six random sites. But on the Moon we have a much clearer view because we don't have the oceans, we don't have the soil cover, and we have very good photography over the surface.

"There were probably some areas of lava flows which may be a little younger than the ones we sampled -so there were these sorts of minor problems. I think we basically already have a very good idea of the overall composition of the lunar surface, the ages and so on. Of course it would be very nice to have extra data on these, I mean there were a lot of second order questions which could be asked, but the first order questions have, in fact, been answered."

HL: "With no atmosphere or weather, would the material on the top of the mountains be the same as the material in the bottom of the valleys?"

Taylor: "More or less. There's not much way of moving material down the slope except during large impacts, when stuff is thrown around and there are examples on the Moon of boulders which have rolled downhill probably due to shaking of the Moon by a large impact. Some of the large boulders which Jack Schmitt looked at in Apollo 17 had come down the hillside. Even though the Apollo 17 site was hundreds of kilometres away from Tycho, the big impact crater, there was a probable landslide caused at the Apollo 17 site by one of the big boulders thrown out from Tycho hitting the top of the hill and causing a landslide in the valley. So there is a certain amount of this goes on, but nothing like the general erosion that goes on all the time on the Earth."

HL: "While you mention Tycho – I suppose cratering has been taking place almost since the formation of the Moon. Is it true that Copernicus and Tycho ring craters were formed more recently?"

Taylor: "Yes they are much younger. In fact Copernicus is used as one of the time markers on the Moon. It formed probably about a billion years ago. Very much later than the big bombardment responsible for most of the big lunar craters in the highlands which ceased about over 4 billion years ago. Copernicus is only about 1 billion years old, and Tycho is probably much younger than that, only about 100 million years old. So that would be during the Jurassic period on the Earth. It's interesting that Tycho is the last big lunar crater – it's about 95 kilometres in diameter – and it must have been formed by something at least 10 kilometres in diameter hitting the Moon. It gives spectacular rays all over the Moon which you can see through field glasses. So it's a very spectacular event and rays from Tycho explosion were thrown right round the Moon. It's interesting that this is not that far away in time from the impact at the Cretaceous Tertiary boundary on the Earth about 65 million years ago. This is the one that probably did in the dinosaurs and that was probably a 10-kilometre object hitting the Earth. So there are these objects floating around, although fortunately they don't hit us very often."

HL: "When did scientists find that tektites no longer came from the Moon? I remember we used to talk quite a bit about tektites and before Apollo we had quite a few in Australia. I remember there was a little exercise to go and find them in the deserts."

Taylor: "Well it is fair to say there was a raging controversy about tektites right through the '60s in which I was heavily involved and I still have a number of scars on me from those arguments because I was on the terrestrial side of the fence. It was a battle fought with great ferocity, particularly since we knew lunar samples were coming back, and the first sample analysis I think was sufficient to dispose of the hypothesis that tektites came from the Moon. As one of my colleagues said, the lunar origin of tektites which had been a stimulating hypothesis for 75 years died on July the 20th 1969 from a massive overdose of lunar data."

HL: "So what effect, if any, does the solar wind have on the lunar surface?"

Taylor: "It impinges almost directly on the Moon – there being no atmosphere or magnetic field on the Moon – so the solar wind comes straight in and impacts on the surface, except of course when the Moon is shielded by the Earth's magnetosphere. So the lunar soil soaks up quite large quantities of hydrogen, helium and rare gases from the solar wind, and this was one of the surprises of the Apollo missions. When the rare gas people got hold of the samples, they found these soils sitting on the surface had very large amounts of hydrogen and helium in them, and other rare gases. This was somewhat of a surprise."

HL: "Why is the Moon evenly bright, when viewed from the Earth, over the whole sphere?"

Taylor: "Well this is one of the ancient puzzles about the Moon, because if you have a smooth sphere and you illuminate it you know it's going to be bright in the middle and darker toward the edges."

HL: "The Earth's like that?"

Taylor: "The Earth's like that, so that when you look at the Moon it could almost be a flat plate in the sky. Because it's bright right across, and the reason for this is it's covered with rubble from these impacts. It's like a multitude of reflectors – a bit like the reflectors on a bicycle. There are all sorts of crystals and rocks and boulders and angles that go right across the surface and you can find enough of them which will be looking directly back at the Earth. So you get this even illumination right across the surface – it was figured out that the Moon must have rubble on it in the late '50s. I think one of the people working on this was the late Professor Jaeger at the Research School of Earth Sciences at ANU. This was the sort of problem you could work on at that time."

HL: "Can you tell us something about the *mascons* which affected a lot of the missions?"

Taylor: "Yes. This again was one of those unexpected things that was discovered on the Moon. *Mascons* are variations in the density of the near lunar surface, and this, of course, affected the trajectory of the spacecraft – and one of the reasons Apollo 11 came down off course was that the trajectory was being affected by these variations in the gravity field. These occur over the dark patches of the Moon which are the so-called lunar maria – the dark circular patches where lava which has come up and filled these large basins. So you get these mass concentrations near the surface.

"One of the causes of the *mascons* are the loading on the surface by these lava flows filling these large impact basins. The other one is that when these big impact basins form – these things are the size of France – they are formed by 100-kilometre objects hitting the Moon. This has the effect of a large blow from above. This has a rebound affect, and part of the more dense lunar matter rises up towards the surface. In some of the craters you get a central peak resulting from this. In the case of the large basins, you get this central core of denser material which has risen up, and this, plus the lava which has flooded into the basin, causes an excess gravity field localised over this basin, and this distorts the orbit of the spacecraft."

HL: "Dr Taylor, are we still working on the analysis of the lunar samples?"
Taylor: "Well there is a quite a body of people still working on the samples, mostly in America, and there is little bit of work still going on in Australia. Because of the results of the Voyager and other missions, a lot of related work on meteorites and the early solar system, has blossomed out into an investigation of the whole of the solar system. The Moon is a key issue because we have the samples and the dates from them, which we don't have from any other planetary body – except we do have some meteorite samples from Mars which have been interpreted in the light of what we learnt on the Moon so the Moon really acted as a Rosetta Stone.

"I am now working on basic models for the origin of the solar system. What started out as lunar work has now expanded out into the whole solar system. This shows up at the annual conferences we have at Houston where now only about ten percent of the meeting is devoted to lunar samples, and the rest of the meeting is on the solar system."

HL: "How do we get meteorites from Mars?"

Taylor: "They are actually fist sized chunks of rock. They are proper meteorites and these have been knocked off Mars by meteorite collisions with Mars."

HL: "Enough to overcome the gravity and atmosphere of Mars?"

Taylor: "Yes. There was a lot of discussion to begin with and some theoretician said it was impossible to get them off Mars without destroying them because of the shock pressure but it turns out when you do this – when you hit Mars and blow a crater – some of the rocks get thrown off which are not particularly shocked – some of the rocks in the edge get into Earth crossing orbit.

"These samples have been known on the Earth as meteorites for quite a long time and they were known to be somewhat funny in comparison with the other meteorites. Mostly they were quite a lot younger and the definitive piece of evidence was that they contained some trapped gases which matched the atmospheric gas composition recorded by the Viking landers on Mars. This was the decisive bit of evidence – there was a lot of circumstantial evidence – but the fact that the rare gases neon, krypton, and xenon isotopic compositions matched what the Viking Landers measured on Mars – it is unlike the Earth's atmosphere – this was the decisive piece of evidence."

HL: "Where were they found?"

Taylor: "Some in Antarctica, but some scattered over the Earth. They were known to be somewhat strange meteorites, some had been in the collection for a long time. Then we've also found half a dozen meteorites from the Moon in Antarctica and one in Western Australia. These are now identified as lunar since we have the lunar samples to identify them, otherwise we would again have a mystery."

HL: "Are the moonrock samples helping you with your solar system research?"

Taylor: "Oh, yes, mainly because of the ages and dates … and the chemistry."

HL: "Twenty-five years on, and you have had some time to think about all this – how do you see the Apollo missions, and what we have found from the lunar samples – taking us into the future?"

Taylor: "You know, I sometimes think when you've spent most of your life working on one subject you should eventually come to some sort of conclusions about it. I've been struck by the role of chance in all these matters. The origin of the Moon was very much a chance event. It turns out that the conditions under which the collision must have occurred – it had to be the right velocity, the right angle, the right mass and so on – although there are some plusses and minuses, nevertheless if the object had hit the Earth head on we would not have formed the Moon.

"Then when you come a bit closer to the event that wiped out the dinosaurs and about seventy or eighty percent of the species on the Earth at the Cretaceous Tertiary boundary 65 million years ago was due to the chance impact of an asteroid – if the asteroid had missed we would not be sitting here discussing this – there would probably be some dinosaur or their derivatives having this conversation. This puts a chance premium on human development. But even when you look at the solar system and the fact that we have the 24-hour rotation of the Earth is probably a consequence of this big collision that formed the Moon and tilted the Earth, and gives us the seasons.

"Venus, in contrast, is rotating very slowly backwards, and has very little tilt on it. Mars is again turning at about 24 hours and has about a 25° tilt, although Mars wobbles about quite a bit, whereas the Earth is locked into this 24° tilt by the stabilising influence of the Moon. These are all chance events. It turns out when you look at the formation of the giant planets, Jupiter is quite difficult to form, very fine timing is required to form a planet like Jupiter. If you don't have a big planet out there to intercept most of the asteroids, meteorites and comets coming in we would have a much higher bombardment rate.

"It seems to me there are so many chance events involved in the formation of the solar system and Earth, that it is very unlikely that there would be any clones of the Earth anywhere else. As a result of these chance events, the prospect of finding life elsewhere, in the way that we know it, on a planet like the Earth, is remote. The philosophical conclusion is that we ought to take much better care of ourselves than we do. If people realise that we may be unique in the universe they may well behave somewhat better than they do."

Analysing the First Apollo 11 Sample of Lunar Rock

Dr Taylor describes the moment of the analysis of the first Apollo 11 sample: "The Lunar landing was on July 20, 1969 – on July 26, the first sample box was opened. The dust coated rocks resembled charcoal briquettes used for barbecues. By 12:28 pm on July 28 the first sample, No. 10084, was prepared and loaded into the cup shaped electrode, which formed the anode of a direct current arc. The excitement of the moment and the tendency of the lunar dust to adhere to everything made this routine operation very difficult. The 10-amp direct current arc was struck. It was designed to volatilise refractory silicates and produce atomic emission spectra, characteristic of the elements present. The sample flared red, indicative of calcium. A little later (the burn took two and a half minutes) a white fringe to the flame suggested that titanium might be a major component."

A Brief Summary of the Apollo Geology Results

Dr Ross Taylor says, "Each mission produced its surprises. Apollo 11 provided unusual chemistry and ancient rocks. Apollo 12 revealed the existence of an extremely fractionated rock type, labelled KREEP (K for potassium, Rare Earth Elements, Phosphorus).

Apollo 14 yielded a plethora of breccias. A peculiar green glass of primitive composition appeared from the Apollo 15 site. Apollo 16, expected by some people to sample volcanic rocks found none, while Apollo 17, looking for young cinder cones, found old orange glasses. The moral, reinforced by our observations on Mars, is that geological processes are different on other planets, and that the value of terrestrial analogies and experience is limited."

Figure A.4. Dr Ross Taylor operating the emission spectrograph under quarantine conditions in the Lunar Receiving Laboratory at the NASA Manned Spacecraft Center in Houston, Texas, during the first analysis of the Apollo 11 samples in July 1969. Taylor says, "With the data from this spectrograph I was able to carry out a complete chemical analysis of the samples given to me."

Table A.1. Major element composition of soils from the Apollo landing sites

Chemical composition values given in wt.%.

APOLLO	11	12	14	15	16	17
Silica	41.3	46.0	47.3	46.0	45.0	40.0
Titanium oxide	7.5	2.8	1.6	1.1	0.29	8.3
Alumina	13.7	12.5	17.8	18.0	29.2	12.1
Iron oxide	15.8	17.2	10.5	11.3	4.2	17.1
Magnesium oxide	8.0	10.4	9.6	10.7	3.9	10.7
Calcium oxide	12.5	10.9	11.4	12.3	17.6	10.8
Sodium oxide	0.41	0.48	0.70	0.43	0.43	0.39
Potassium oxide	0.14	0.26	0.55	0.16	0.06	0.09
Manganese oxide	0.21	0.22	0.14	0.15	0.06	0.22
Chromium oxide	0.29	0.41	0.20	0.33	0.08	0.41
Totals	99.8	101.0	99.8	100.5	100.8	100.5

Source: Planetary Science: A Lunar Perspective, Stuart Ross Taylor (1982).

Table A.2. Times of major lunar events

Origin of the Moon	4,570 million years ago
Melting of the outer 400 kilometres	4,300 million years ago
End of the cataclysmic meteorite bombardment	3,860 million years ago
Copernicus crater	800 million years ago
Tycho crater	97 million years ago

A Quiz

Some of your questions on space answered. The answers will be found on the page number in brackets.

1 Where and when was the first sustained flight by a hot air balloon tried? (page 4)

2 Who launched the first successful liquid-fuelled rocket? (page 6)

3 Who was the first person to fly using a rocket for propulsion? (page 7)

4 Where does space officially begin? (page 11)

5 What was the first man-made object into the vacuum of space? (page 10)

6 When was the first radar signal bounced off the Moon detected on Earth? (page 11)

7 Who was the first pilot to break through the sound barrier? (page 12)

8 Why did the Russians choose dogs and the Americans chimpanzees to trial conditions for life in space? (pages 17 and 64)

9 How did NASA decide to go to the Moon? (page 20)

10 Which spacecraft first left the Earth's gravitational field? (page 19)

11 Which spacecraft first hit Moon? (page 20)

12 When is the Moon upside down? (page 20)

13 Which spacecraft sent the first pictures of the far side of the Moon? (page 20)

14 Why did Russia choose spherical re-entry spacecraft and American truncated cones? (page 31)

15 What qualifications did the original American astronauts need? (page 33)

16 How many 'g's' can a human withstand? (page 35)

17 What is the difference between the "orbit" and "revolution" of a spacecraft around a planet? (page 53)

18 Why did the Gemini spacecraft use an ejection seat for the pilots' safety, while the Mercury and Apollo Projects used rocket escape towers? (page 92)

19 Who was the first person to walk in space? (page 96)

20 When was a computer first used in space? (page 100)

21 Which was the first manned mission supported by the Houston Mission Control Center, later to become the Johnson Space Center? (page 100)

22 Which spacecraft first soft-landed on the Moon's surface? (page 119)

23 Who did the first hard docking between two spacecraft in space? (page 119)

24 Who was the first person to walk completely around the world in space? (page 126)

25 Who was the first astronaut to fall asleep in the middle of a spacewalk? (page 133)

26 Which mission tried the first artificial gravity experiment? (page 133)

27 Why was the big Apollo moonrocket named "Saturn"? (page 149)

28 How were Armstrong, Aldrin and Collins, chosen to be the crew for the Apollo 11 first Moon landing mission? (pages 193 and 209)

29 Who was the first space traveller to die in space? (page 160)

30 How did the first man in space, Yuri Gagarin, lose his life? (page 169)

31 Who was the only American astronaut to fly in the Mercury, Gemini and Apollo missions? (page 175)

32 Who were the first humans to leave the Earth and its immediate environment and head out into space towards the Moon? (page 178)

33 When was the first crew transfer between two spacecraft? (page 194)

34 Looking from the North Pole, why did the Apollo missions to the Moon always fly clockwise around the Moon? (page 210)

35 Which part of the Apollo mission was regarded by Houston Flight Controllers as the most dangerous? (page 214)

36 How many jumbo jet aircraft does it take to equal the power of the Saturn V moon-rocket? (page 214)

37 How many tons of fuel per second did the Saturn V rocket consume to get off the ground? (page 215)

38 Which tracking station relayed the picture of Armstrong's first step on the Moon to the world? (page 233)

39 What sort of still camera equipment did Armstrong use on the Moon's surface? (page 240)

40 Who saw the first eclipse of the Sun by the Earth? (page 264)

41 What fault caused the explosion on the Apollo 13 flight? (page 268)

42 What would happen to you if you were suddenly exposed to the vacuum of space? (page 301)

43 Who drove the first vehicle on the Moon? (page 309)

44 Who found the so-called Apollo "Genesis Rock"? (page 312)

45 Many people could probably figure out Shepard was the oldest astronaut to walk on the Moon. Who was the youngest, and how old was he? (page 316)

46 What is the speed record on the Moon's surface? (page 326)

47 What was the biggest impact on the Moon ever recorded by the Apollo Lunar Science Experiment Package seismometers? (page 328)

48 Who did the first automotive repair on the Moon? (page 334)

49 Is the totally jet-black cloudless sky of the Moon oppressive, like a room with a black ceiling? (page 336)

50 Who was the last man to leave the Moon's surface in the Apollo Program? (page 336)

51 Why did Apollo 13 get cold when in trouble, and Skylab get hot? (page 345)

52 How many sunsets a day did the astronauts in Skylab see? (page 355)

53 What effect does travelling in space and weightlessness have on the human body? (page 366)

54 Do astronauts suffer motion sickness out in space? (page 367)

55 Without gravity to pull blood away from the head, does the brain work better in the weightless environment of space? (page 367)

56 How much did the Apollo Program cost the American taxpayer? (Appendix 1, page 389)

57 What happened to all the moonrock samples that were brought back? (Appendix 1, page 389)

58 When did the Apollo scientific experiments left behind by the astronauts, ALSEP, stop sending information back to earth? (Appendix 1, page 389)

59 Does the lunar soil have a smell? (Appendix 2, page 394)

60 Why do shock waves from a heavy object impacting the Moon last for up to four hours? (Appendix 2, page 395)

61 From earth we see the Moon evenly-lit all over, like a flat plate. As the Moon is a sphere, why isn't there a shiny spot where the Sun strikes the surface, as would be the case for a billiard ball? (Appendix 2, page 402)

62 Who was the first human buried on the Moon? (page 385)

References

Voice exchanges between the spacecraft and ground are printed in purple in the text, and have generally been transcribed from tapes recorded during the missions. Apart from the personal communications and interviews, significant references and quotations are from the following sources:

1. "HISTORY OF ROCKETRY AND SPACE TRAVEL"
   By Wernher von Braun and Frederick I. Ordway III
   Copyright © 1969, 1966 by Wernher von Braun, Frederick I. Ordway III, and Harry H-K Lange.
   Thomas Y. Crowell Company, Inc., Publishers.

2. "THE NATIONAL AIR & SPACE MUSEUM"
   by C.D.B. Ryan.
   Second Edition. © The Smithsonian Institution,
   Washington, DC. (1988, Harry N. Abrams, Inc.)

3. "ORDERS OF MAGNITUDE"
   A History of NACA & NASA: 1915–1980.
   by Frank W. Anderson.
   1981. NASA SP-4403. NASA, Washington, DC.

4. "VENTURE INTO SPACE"
   Early Years of Goddard Space Flight Center.
   by Alfred Rosenthal.
   1968, NASA SP-4301, NASA, Washington, DC.

5. Excerpts from "1928–1929 Forerunners of the Shuttle: The von Opel Flights."
   Spaceflight. Volume 21, February 1979, p. 75.
   With kind permission from the author, Frank H. Winter,
   Curator of Rocketry, National Air & Space Museum,
   Smithsonian Institution, Washington, DC.

6. Information supplied by the Air & Space Museum library,
   Smithsonian Institution, Washington, DC.

7. From discussions with Mr Frank Winter, Curator of Rocketry, National Air & Space Museum, Smithsonian Institution, Washington DC with reference to Willy Ley.

8. "Soviet Space – The First Twenty Years."
   By Professor Gregori Tokaty.
   New Scientist Magazine. October 6, 1977.

9. "THIS NEW OCEAN"
   A History of Project Mercury.
   by Lloyd Swenson, Jr, James Grimwood, and Charles Alexander.
   1966. NASA SP-4201. NASA, Washington, DC.

10. Excerpts reprinted with kind permission of the Society from:
    The Society of Experimental Test Pilots' Proceedings, 1975.
    Lancaster, CA. Part of a lecture by Captain Charles E. Yeager.

11. "The World in Space"
    The Story of the International Geophysical Year.
    By Alexander Marshack.
    Copyright ☐ 1958 Alexander Marshack.
    Published by Dell Publishing Co. Inc. New york 17, NY.

12. Histories of the: Space Tracking & Data Acquisition Network STADAN
                     Manned Space Flight Network MSFN
                     NASA Communications Network NASCOM
    by William R. Corliss.
    June 1974. NASA CR-140390. NASA, Washington, DC.

13. "BEYOND THE ATMOSPHERE"
    The Early Years of Space Science.
    By Homer E. Newall
    1980. NASA SP-4211. NASA, Washington, DC.

14. Information from "TO A ROCKY MOON"
    A Geologist's History of Lunar Exploration.
    By Don E. Wilhelms.
    © 1993 The Arizona Board of Regents.
    The University of Arizona Press, Tucson and London, UK.

15. "LUNAR IMPACT"
    A History of Project Ranger.
    by R. Cargill Hall.
    1977, NASA SP-4210, NASA, Washington, DC.

16. "PROJECT MERCURY – A Chronology"
    by James Grimwood.
    1963. NASA SP-4001, NASA Washington, DC.

17. "APOLLO EXPEDITIONS TO THE MOON"
    Edited by Edgar M. Cortright.
    1975, NASA SP-350, NASA, Washington, DC.

18. "A HISTORY OF THE DEEP SPACE NETWORK"
    by William R. Corliss
    May 1, 1976. NASA CR-151915. NASA, Washington, DC.

19. ROUNDUP Newspaper NASA Manned Spacecraft Center, Houston.
    Volume 8 No 25 October 3, 1969.

20. Excerpts reprinted with the kind permission of Macmillan Publishing Company from the book "INTO ORBIT",
    by the Seven Astronauts of Project Mercury, introduced by John Dill.
    © 1962 Time Inc.
    Originally published by Cassell and Co., Ltd., London.

21. "GEMINI – A Personal Account of Mans' Adventure into Space"
    by Virgil Grissom.
    1968. World Book, New York, NY.

22. MERCURY PROJECT SUMMARY,
    including results of the Fourth Manned Orbital Flight.
    May 15 and 16, 1963. NASA SP-45. NASA, Washington, DC.

23. NASA Networks "TECHNICAL INFORMATION BULLETIN"

24. "LUNAR ORBIT RENDEZVOUS"
    Transcript of news conference.
    1962, US Government Printing Office
    Publication O-654534

25. Results of the Second United States Manned Orbital Space Flight.
    May 24, 1962. NASA SP-6. NASA, Washington, DC.

26. "Minolta Messenger"
    No 7. Published by the Corporate Communications Division of the Minolta Camera Company Limited. 30,
    2-chrome, Azuchi -Machi, Higashi-ku, Osaka, Japan

27. Results of the First United States Manned Orbital Space Flight.
    February 20, 1962. NASA Publication. Washington, DC.

28. The Australian Senate Parliamentary Debate (Hansard.)

29. "DEKE!"
    US Manned Space: From Mercury to the Shuttle.
    By Donald K. Slayton with Michael Cassutt.
    1994. Forge Book published by Tom Doherty Associates, Inc.
    175 Fifth Avenue, New York, NY. 10010.
    Copyright © 1994.

30. Results of the Third United States Manned Orbital Space Flight.
    October 3, 1962. NASA SP-12. NASA, Washington, DC.

31. From: "A WALK IN SPACE" NASA Booklet.
    No author, date, or other identification.
    US Government Publication.

32. "ON THE SHOULDERS OF TITANS"
    By Barton Hacker and James Grimwood.
    1977. NASA SP-4203. NASA, Washington, DC.

33. "CARRYING THE FIRE"
    by Michael Collins.
    1974, Farrar, Straus & Giroux, Inc. New York, NY.
    Copyright © 1974 by Michael Collins.

34. "BEYOND SOUTHERN SKIES"
    By Peter Robinson.
    1992. Cambridge University Press, Cambridge, United Kingdom.

35. "UNCOVERING SOVIET DISASTERS"
    By James E. Oberg.
    1988. Random House, New York, NY.

36. Excerpt reprinted from "Engineering Aspects of the Lunar Landing"
The Gardner Lecture, MIT.
by Dr Robert Seamans and Mr Neil Armstrong, 1994.
Supplied to the author by Neil Armstrong.

37. Personal communication with Mr Christopher Kraft, Director, Johnson Space Center.

38. "EARTHBOUND ASTRONAUTS"
The builders of Apollo-Saturn.
By Beirne Lay. Jr.
Copyright □ 1971 by Beirne Lay, Jr.
Prentice-Hall, Inc., Englewood Cliffs, NJ.

39. "Aviation Week & Space Technology Magazine"
May 19, 1969.

40. ROUNDUP Newspaper NASA Manned Spacecraft Center Houston.
Volume 7 No 25.
September 27, 1968.

41. "PARKES: Thirty Years of Radio Astronomy"
Bolton, J.G. (1994), PARKES and the Apollo Missions,
Edited by D.E. Goddard and D.K.Milne.
CSIRO Publications, Melbourne, Victoria.
Excerpts, with kind permission from the authors.

42. NASA ACTIVITIES
July 1979. "The Way it Was."

43. "RETURN TO EARTH"
Colonel Edwin E. 'Buzz' Aldrin, Jr., with Wayne Warga.
Copyright © 1973 by Aldrin-Warga Associates.
Published by Random House, New York, NY.
Excerpts reprinted by kind permission of Aldrin-Warga Associates.

44. The Society Of Experimental Test Pilots' Proceedings, 1969.
Lancaster, CA. Selections from a lecture by Neil Armstrong.
Excerpts reprinted with the kind permission of Neil Armstrong.

45. "Aviation Week & Space Technology Magazine"
August 11, 1969.

46. "Aviation Week & Space Technology Magazine"
December 8, 1969.

47. "APOLLO EXPEDITIONS TO THE MOON"
Edited by Edgar M. Cortright.
1975, NASA SP-350, NASA, Washington, DC.

48. Australian House of Representatives Parliamentary Debate (Hansard)

49. "GODDARD NEWS"
Volume 18 No 2. May 11, 1970.

50. Information from a paper on ESP by Edgar Mitchell.

51. Excerpts from "TO RULE THE NIGHT"
James B. Irwin with William Emerson, Jr.
1973, Hodder & Stoughton, London.

52. "What Is It Like To Walk on the Moon?"
By David R. Scott.
NATIONAL GEOGRAPHIC Magazine
Vol 144, No 3. September 1973
Excerpts reprinted with kind permission of the National Geographic Society, Washington, DC.

53. "THE LUNAR ROVING VEHICLE"
By Eugene G. Cowart, LRV Chief Engineer.
A Boeing Company Publication. Undated.

54. "MOONWALKER"
by Charlie and Dotty Duke.
Copyright © 1990 by Charlie and Dotty Duke.
Oliver-Nelson Books, Nashville.
Excerpts reprinted with kind permission from Charlie and Dotty Duke.

55. "SKYLAB – Our First Space station"
Edited by Leland F. Belew.
1977. NASA SP-400.
National Aeronautics and Space Administration, Washington, DC.

56. Hasselblad brochure written by Evald Karisten, 1974

57. Guinness Book of Records.
Copyright © 2000 Guinness World Records Ltd.

58. The Last Man on the Moon
Eugene Cernan with Don Davis
© Eugene Cernan and Don Davis 1999
St Martin's Press, 175 Fifth Avenue, New York, NY 10010

59. The Illustrated Encyclopedia of Space Technology
Principal Author Kenneth Gatland Editor Philip de Ste. Croix
© Salamander Books Ltd 1981
27 Old Gloucester Street, London. WC1N 3AF

Index and Glossary

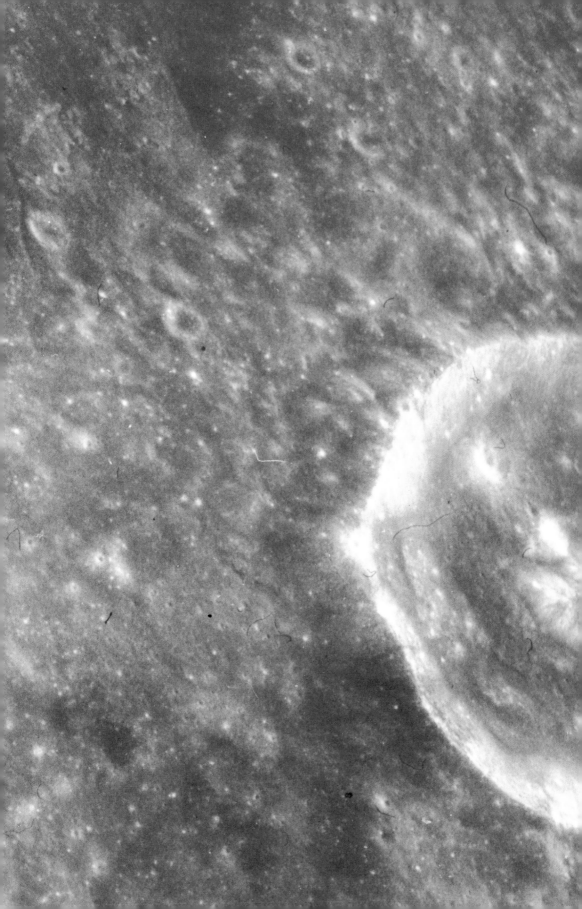